A&P TECHNICIAN
GENERAL
2000
STUDY GUIDE

© Jeppesen Sanderson Inc., 1996, 1997, 1998, 2000
All Rights Reserved
55 Inverness Drive East, Englewood, CO 80112-5498
ISBN 0-88487-204-1

PREFACE

Thank you for purchasing this *Aviation Mechanic General Knowledge Test Study Guide*. This study guide will help you understand the answers to the test questions so you can take the FAA computer exam with confidence. It contains all FAA Aviation Mechanic General Knowledge test questions. Included are the correct answers and explanations, along with study references. Explanations of why the other choices are wrong have been included where appropriate. Questions are organized by topic, with explanations conveniently located adjacent to each question. Figures identical to those on the FAA test are included, plus our unique sliding mask for self-testing. Please note that this Study Guide is intended to be a supplement to your instructor-led maintenance training, not a stand-alone learning tool.

THE JEPPESEN SANDERSON TRAINING PHILOSOPHY

Maintenance training in the developing years of aviation was characterized by the separation of academics from maintenance training on the aircraft. For years, the training of theory and practice were not integrated. There were lots of books on different subjects, written by different authorities, which produced a general lack of continuity in training material. The introduction of **Jeppesen Sanderson Training Products** changed all this. Our proven, professional, integrated training materials include extensive research on teaching theory and principles of how people learn best and most efficiently. Effective instruction includes determining objectives and completion standards. We employ an important principle of learning a complex skill using a step-by-step sequence known as the **building block principle**. Another important aspect of training is the principle of **meaningful repetition**, whereby each necessary concept or skill is presented several times throughout the instructional program. Jeppesen training materials incorporate these principles in our textbooks, exercises, exams, and this Study Guide. When these elements are combined with an instructor's class discussion and the skills learned in the shop, you have an ideal integrated training system, with all materials coordinated.

Observation and research show that people tend to retain 10% of what they read, 20% of what they hear, 30% of what they see, and 50% of what they both hear and see together. These retention figures can be increased to as high as 90% by including active learning methods. Video and textbook materials are generally considered passive learning methods. Exercises, stage exams, student/instructor discussions, and skills in the shop are considered to be active learning methods. Levels of learning include rote, understanding, application, and correlation. One of the major drawbacks with test preparation courses that concentrate only on passing the test is that they focus on rote learning, the lowest level of learning employed in a teaching situation. Students benefit from Jeppesen's professional approach through standardized instruction, a documented training record, increased learning **and** increased passing rates. Our materials are challenging and motivating, while maximizing knowledge and skill retention. Thousands of technicians have learned aviation maintenance using our materials, which include:

MANUALS — Our training manuals contain the answers to many of the questions you may have as you begin your training program. They are based on the **study/review** concept of learning. This means detailed material is presented in an uncomplicated way, then important points are summarized through the use of bold type and color. The best results can be obtained when the manual is studied as an integral part of the coordinated materials. The manual is the central component for academic study.

SUPPORT COMPONENTS — Supplementary items include an exercise book, stage and final exams, FAR handbook, Aircraft Technical Dictionary, and a Standard Aviation Maintenance Handbook. In addition, Jeppesen offers in-depth guidebooks on individual aviation maintenance subject areas. Jeppesen Sanderson's training products are the most comprehensive technician training materials available. They help you prepare, in conjunction with your instructor, for the FAA exam and practical test; and, more importantly, they help you become a more proficient and safer technician.

You can purchase our products and services through your local Jeppesen dealer. For product, service, or sales information call **1-800-621-JEPP, 303-799-9090, or FAX 303-784-4153**. If you have comments, questions, or need explanations about any component of our Technician Training System, we are prepared to offer assistance at any time. If your dealer does not have a Jeppesen catalog, please request one and we will promptly send it to you. Just call the above telephone number, or write:

    Marketing Manager, Training Products
    Jeppesen Sanderson, Inc.
    55 Inverness Drive East
    Englewood, CO 80112-5498

Please direct inquiries from Europe, Africa, and the Middle East to:

    Jeppesen & Co., GmbH
    P. O. Box 1564
    63235 Neu-Isenburg
    Germany
    Tel: 0 61 02/50 70
    Fax: 0 61 02/50 79 99

v

TABLE OF CONTENTS

PREFACE iii

INTRODUCTION ix

CHAPTER 1   **Mathematics** 1-1
   Section A   Arithmetic 1-1
   Section B   Algebra 1-10
   Section C   Geometry and Trigonometry 1-12

CHAPTER 2   **Physics** 2-1
   Section A   Matter and Energy 2-1
   Section B   Work, Power, Force, and Motion 2-1
   Section C   Gas and Fluid Mechanics 2-3
   Section D   Aerodynamics 2-5
   Section E   High-Speed Aerodynamics 2-7
   Section F   Helicopter Aerodynamics 2-7

CHAPTER 3   **Basic Electricity** 3-1
   Section A   Theory and Principles 3-1
   Section B   Direct Current 3-7
   Section C   Batteries 3-16
   Section D   Alternating Current 3-19
   Section E   Electron Control Devices 3-24
   Section F   Electrical Measuring Instruments 3-28
   Section G   Circuit Analysis 3-28

CHAPTER 4   **Electrical Generators and Motors** 4-1
   Section A   DC Generators 4-1
   Section B   Alternators 4-1
   Section C   Motors 4-1

CHAPTER 5   **Aircraft Drawings** 5-1
   Section A   Types of Drawings 5-1
   Section B   Drawing Practices 5-9
   Section C   Charts and Graphs 5-18

CHAPTER 6   **Weight and Balance** 6-1
   Section A   Weighing Procedures 6-1
   Section B   Shifting the CG 6-6
   Section C   Helicopter Weight and Balance 6-9

CHAPTER 7   **Aircraft Structural Materials** 7-1
   Section A   Metals 7-1
   Section B   Nonmetallic Materials 7-4

CHAPTER 8	**Aircraft Hardware**	**8-1**
Section A	Aircraft Rivets	8-1
Section B	Aircraft Fasteners	8-1
CHAPTER 9	**Hand Tools and Measuring Devices**	**9-1**
Section A	Hand Tools	9-1
Section B	Measuring and Layout Tools	9-1
CHAPTER 10	**Fluid Lines and Fittings**	**10-1**
Section A	Rigid Fluid Lines	10-1
Section B	Flexible Fluid Lines	10-5
CHAPTER 11	**Nondestructive Testing**	**11-1**
Section A	Visual Inspections	11-1
Section B	Electronic Inspections	11-7
CHAPTER 12	**Cleaning and Corrosion**	**12-1**
Section A	Aircraft Cleaning	12-1
Section B	Types of Corrosion	12-3
Section C	Corrosion Detection	12-5
Section D	Treatment of Corrosion	12-5
CHAPTER 13	**Ground Handling and Servicing**	**13-1**
Section A	Shop Safety	13-1
Section B	Flight Line Safety	13-1
Section C	Servicing Aircraft	13-6
CHAPTER 14	**Maintenance Publications, Forms, and Records**	**14-1**
Section A	Maintenance Publications	14-1
Section B	Forms and Records	14-8
CHAPTER 15	**Mechanic Privileges and Limitations**	**15-1**
Section A	The Mechanic Certificate	15-1
APPENDIX 1		**C-1**
APPENDIX 2		**Q-1**

INTRODUCTION _____

The *Aviation Mechanic General Knowledge Test Study Guide* is designed to help you prepare for the FAA Aviation Mechanic General Knowledge computerized test. It covers FAA exam material that applies to general knowledge related to aircraft maintenance.

We recommend that you use this Study Guide in conjunction with the Jeppesen Sanderson A&P Technician General Textbook. The Study Guide is organized along the same lines as the General Textbook, with 15 chapters and distinctive sections within most chapters. Questions are covered in the Study Guide in the same sequence as the material in the manual. References to applicable chapters and pages in the various manuals are included along with the answers.

Within the chapters, each section contains a brief introduction and a list of questions in that section. FAA exam questions appear in the left-hand column of the Study Guide, while answers are in the right-hand column. The first line of the answer for each question is in bold type with the question number, the answer, and the page where the question is covered in the A & P Technician General Textbook. There is also an abbreviation for an FAA or other authoritative source document.

Example: 8324. Answer B. JSGT 13-28 (AC 65-9A)

Next is a brief explanation of the correct answer, followed by an explanation of why the other answers are wrong. In some cases the incorrect answers are not explained. Examples include instances where the answers are calculated, or when the explanation of the correct answer obviously eliminates the wrong answers.

Abbreviations used in the Study Guide are as follows:

AC	—	Advisory Circulars
AC 65-9A	—	Airframe and Powerplant Mechanics General Handbook
AC 65-12A	—	Airframe and Powerplant Mechanics Powerplant Handbook
AC 65-15A	—	Airframe and Powerplant Mechanics Airframe Handbook
ASTM	—	American Society for Testing and Materials
FA 150	—	Airborne Digital Logic Principles
FAR	—	Federal Aviation Regulation
JSGT	—	Jeppesen Sanderson General Textbook
JSAT	—	Jeppesen Sanderson Airframe Textbook
JSPT	—	Jeppesen Sanderson Powerplant Textbook

Since the FAA does not provide answers with their test questions, the answers in this Study Guide are based on official reference documents and, in our judgement, are the best choice of the available answers. Some questions which were valid when the FAA Computerized Test was originally released may no longer be appropriate due to changes in regulations or official operating procedures. However, with the computer test format, timely updating and validation of questions is anticipated. Therefore, when taking the FAA test, it is important to answer the questions according to the latest regulations or official operating procedures.

One appendix from the FAA test materials is included in the back of the Study Guide. This is Appendix 1 which includes Subject Matter Knowledge Codes and reference materials. Appendix 2 in the Study Guide consists of a numerical listing of all the questions. Included in this listing is a tabulation with the answer and the page number where the question appears in the study guide.

Figures in the Study Guide are the same as those that are used in the FAA Computerized Testing Supplement. These figures, which are referred to in many of the questions, are placed throughout the Study Guide as close as practical to the applicable questions. When a figure is not on the same page or facing page, a note will indicate the page number where you can find that figure.

While good study material is beneficial, it is important to realize that to become a safe, competent technician, you need more than just the academic knowledge required to pass a written test. A certified Airframe and Powerplant Mechanics school will give you the practical shop skills that are indispensible to mechanics working in the field.

WHO CAN TAKE THE TEST

The Aviation Mechanic General Exam is usually taken in conjunction with either the Aviation Mechanic Airframe or Powerplant exam. When you are ready to take one of these FAA computerized tests, you must present either a graduation certificate or certificate of completion from a certificated aviation maintenance technician school, or documentary evidence of practical work experience. For a single rating, you must have at least 18 months of practical experience with the procedures, practices, and equipment generally used in constructing, maintaining, or altering airframes or powerplants. To test for both ratings, you must show at least 30 months of practical experience concurrently performing the duties appropriate to both the airframe and powerplant ratings. Documentary evidence of practical experience must be satisfactory to the administrator.

You also must provide evidence of a permanent mailing address, appropriate identification, and proof of your age. The identification must include a current photograph, your signature, and your residential address, if different from your mailing address. You may present this information in more than one form of identification, such as a driver's license, government identification card, passport, alien residency (green) card, or a military identification card.

HOW TO PREPARE FOR THE FAA TEST

It is important to realize that to become a safe, competent mechanic, you need more than just the academic knowledge required to pass a test. For a comprehensive training program, we recommend a structured maintenance school with qualified instructors. An organized course of instruction will help you complete the course in a timely manner, and you will be able to have your questions answered.

Regardless of whether or not you are in a structured ground training program, you will find this Study Guide is an excellent training aid to help you prepare for the FAA computerized test. The Study Guide contains all of the FAA questions as they are presented in the FAA computerized test format. By reviewing the questions and studying the Jeppesen Sanderson Maintenance Training materials, you should be well equipped to take the test.

You will also benefit more from your study if you test yourself as you proceed through the Study Guide. Cover the answers in the right-hand column, read each question, and choose what you consider the best answer. A sliding mask is provided for this purpose. Move the sliding mask down and read the answer and explanation for that question. You may want to mark the questions you miss for further study and review prior to taking your exam.

The sooner you take the exam after you complete your study, the better. This way, the information will be fresh in your mind, and you will be more confident when you actually take the FAA test.

GENERAL INFORMATION — FAA COMPUTERIZED TESTS

Detailed information on FAA computer testing is contained in FAA Order 8080.6A, *Conduct of Airmen Knowledge Tests Via The Computer Medium*. This FAA order provides guidance for Flight Standards District Offices (FSDOs) and personnel associated with organizations that are participating in, or are seeking to participate in, the FAA Computer-Assisted Airmen Knowledge Testing Program. You also may refer to FAA Order 8300.1, *Airworthiness Inspector's Handbook*, for guidance on computer testing by FAR Part 147 maintenance training schools that hold examining authority.

As a test applicant, you don't need all of the details contained in FAA Orders, but you may be interested in some of the general information about computer testing facilities. A **Computer Testing Designee (CTD)** is an organization authorized by the FAA to administer FAA airmen knowledge tests via the computer medium. A **Computer Testing Manager (CTM)** is a person selected by the CTD to serve as manager of its national computer testing program. A **Testing Center Supervisor (TCS)** is a person selected by the CTM, with FAA approval, to administer FAA airmen knowledge tests at approved testing centers. The TCS is responsible for the operation of the testing center.

CTDs are selected by the FAA's Flight Standards Service. Those selected may include companies, schools, universities, or other organizations that meet specific requirements. For example, they must clearly demonstrate competence in computer technology, centralized database management, national communications network operation and maintenance, national facilities management, software maintenance and support, and technical training and customer support. They must provide computer-assisted testing, test administration, and data transfer service on a national scale. This means they must maintain a minimum of 20 operational testing centers geographically dispersed throughout the United States. In addition, CTDs must offer operational hours that are convenient to the public. An acceptable plan for test security is also required.

WHAT TO EXPECT ON THE COMPUTERIZED TEST

Computer testing centers are required to have an acceptable method for the "on-line" registration of test applicants during normal business hours. They must provide a dual method for answering questions, such as keyboard, touch screen, or mouse. Features that must be provided also include an introductory lesson to familiarize you with computer testing procedures, the ability to return to a test question previously answered (for the purpose of review or answer changes), and a suitable display of multiple-choice and other question types on the computer screen in one frame. Other required features include a display of the time remaining for the completion of the test, a "HELP" function which permits you to review test questions and optional responses, and provisions for your test score on an Airmen Computer Test Report.

On the computerized tests, the selection of questions is done for you, and you will answer the questions that appear on the screen. You will be given a specific amount of time to complete the test, which is based on past experience with others who have taken the exam. If you are prepared, you should have plenty of time to complete the test. After you begin the test, the screen will show you the time remaining for completion. When taking the test, keep the following points in mind:

1. Answer each question in accordance with the latest regulations and procedures. If the regulation or procedure has recently changed, you will receive credit for the affected question. However, these questions will normally be deleted or updated on the FAA computerized tests.

2. Read each question carefully before looking at the possible answers. You should clearly understand the problem before attempting to solve it.

3. After formulating an answer, determine which of the alternatives most nearly corresponds with that answer. The answer chosen should completely resolve the problem.

4. From the answers given, it may appear that there is more than one possible answer; however, there is only one answer that is correct and complete. The other answers are either incomplete or are derived from popular misconceptions.

5. Make sure you select an answer for each question. Questions left unanswered will be counted as incorrect.

6. If a certain question is difficult for you, it is best to proceed to other questions. After you answer the less difficult questions, return to those which were unanswered. The computerized test format helps you identify unanswered questions, as well as those questions you wish to review.

7. When solving a calculator problem, select the answer nearest your solution. The problem has been checked with various types of calculators; therefore, if you have solved it correctly, your answer will be closer to the correct answer than the other choices.

8. Generally, the test results will be available almost immediately. Your score will be recorded on an Airmen Computer Test Report form, which includes subject matter knowledge codes for incorrect answers. To determine the knowledge area in which a particular question was incorrectly answered, compare the subject matter knowledge codes on this report to Appendix 1, Subject Matter Knowledge Codes in this book.

Computer testing designees must provide a way for applicants, who challenge the validity of test questions, to enter comments into the computer. In addition to comments, you will be asked to respond to a critique form which may vary at different computer testing centers. The TCS must provide a method for you to respond to critique questions projected on the computer screen. The test proctor should advise you, if you have complaints about test scores, or specific test questions, to write directly to the appropriate FAA office.

TEST MATERIALS, REFERENCE MATERIALS, AND AIDS

You are allowed to use an electronic calculator for this test. Simple programmable memories, which allow addition to, subtraction from, or retrieval of one number from the memory, are acceptable. Simple functions such as square root or percent keys are also acceptable.

In addition, you may use any reference materials provided with the test. You will find that these reference materials are the same as those in this book.

RETESTING

As stated in FAR section 65.19, an applicant who fails a test may not apply for retesting until 30 days after the date the test was failed. However, the applicant may apply for retesting before the 30 days have expired provided the applicant presents a signed statement from an airman holding the certificate and rating sought by the applicant certifying that the airman has given the applicant additional instruction in each of the subjects failed and that the airman considers the applicant ready for retesting.

WHERE TO TAKE THE FAA TEST

Testing is administered via computer at FAA-designated test centers. As indicated, these CTDs are located throughout the U.S. You can expect to pay a fee and the cost varies at different locations. The following is a listing of the approved computer testing designees at the time of publication of this question bank. You may want to check with your local FSDO for changes.

Aviation Business Services
1-800-947-4228
Outside U.S. (415) 259-8550

Sylvan Prometric
1-800-359-3278
1-800-967-1100
Outside U.S. (410) 880-0880, Extension 8890

CHAPTER 1

MATHEMATICS

SECTION A
ARITHMETIC

Section A of Chapter 1 contains information on the fundamentals of arithmetic. Included are the number system, signed numbers, fractions, scientific notation, powers and roots, percentage, ratio, and proportion. The following FAA Computerized Test questions apply to this section:

8379, 8380, 8382, 8383, 8384, 8385, 8386, 8387, 8388, 8389, 8390, 8391, 8392, 8393, 8409, 8410, 8411, 8412, 8413, 8414, 8415, 8416, 8417, 8418, 8419, 8420, 8421, 8422, 8423, 8424, 8425, 8426, 8427, 8428, 8429, 8430, 8435, 8438.

8379. H01
What power of 10 is equal to 1,000,000?

A — 10 to the fourth power.
B — 10 to the fifth power.
C — 10 to the sixth power.

8380. H01
Find the square root of 1,746.

A — 41.7852.
B — 41.7752.
C — 40.7742.

8382. H01
Find the square root of 3,722.1835.

A — 61.00971.
B — 61.00.
C — 61.0097.

8383. H01
$8{,}019.0514 \times 1/81$ is equal to the square root of

A — 9,108.
B — 9,081.
C — 9,801.

8379. Answer C. JSGT 1-11 (AC 65-9A)
When using scientific notation, you can quickly determine the power of ten by counting the number of zeros. In this problem, 1,000,000 has six zeros which is equal to 10 to the sixth power.

8380. Answer A. JSGT 1-11 (AC 65-9A)
The square root of a number is the root of that number multiplied by itself. You can calculate the square root of 1,746 by using a calculator with a square root function, or by simply multiplying each selection by itself to see which one equals 1,746. The answer is 41.7852.

8382. Answer C. JSGT 1-11 (AC 65-9A)
The square root of a number is the root of that number multiplied by itself. You can calculate the square root of 3,722.1835 by using a calculator with a square root function, or by multiplying each selection by itself to see which one equals 3,722.1835. The answer is 61.0097.

8383. Answer C. JSGT 1-11 (AC 65-9A)
Begin solving this problem by converting 1/81 into a decimal. The decimal equivalent of 1/81 is .012 (1 ÷ 81 = .012). Next, multiply 8,019.0514 by .012, the decimal equivalent of 1/81. The product of these two numbers is 99.0006 (8,019.0514 × .012 = 99.0006). Now, square 99.0006 60 to find the answer 9,802.1 (99.0006^2 = 9,802.1). Answer C is the closest.

8384. H01
Find the cube of 64.

A — 4.
B — 192.
C — 262,144.

8385. H01
Find the value of 10 raised to the negative sixth power.

A — 0.000010.
B — 0.000001.
C — 0.0001.

8386. H01
What is the square root of 4 raised to the fifth power?

A — 32.
B — 64.
C — 20.

8387. H01
The number 3.47×10 to the negative fourth power is equal to

A — .00347.
B — 34,700.0.
C — .000347.

8388. H01
Which alternative answer is equal to 16,300?

A — 1.63×10 to the fourth power.
B — 1.63×10 to the negative third power.
C — 163×10 to the negative second power.

8389. H01
Find the square root of 124.9924.

A — 111.8×10 to the third power.
B — $.1118 \times 10$ to the negative second power.
C — $1,118 \times 10$ to the negative second power.

8390. H01
What is the square root of 16 raised to the fourth power?

A — 1,024.
B — 4,096.
C — 256.

8384. Answer C. JSGT 1-11 (AC 65-9A)
The cube of a number is that number multiplied by itself three times. The answer is 262,144 or $(64 \times 64 \times 64 = 262,144)$.

8385. Answer B. JSGT 1-11 (AC 65-9A)
When working with powers of 10, a negative exponent indicates you must move the decimal to the left. The number of places the decimal should be moved is equivalent to the exponent's value. Therefore, 10^{-6} is equal to .000001.

8386. Answer A. JSGT 1-11 (AC 65-9A)
The square root of four is two. Two multiplied by itself five times equals 32.

$$(\sqrt{4})^5 = (2)^5 = 32$$

8387. Answer C. JSGT 1-11 (AC 65-9A)
When working with powers of 10, a negative exponent indicates you must move the decimal to the left. The number of places the decimal should be moved is equivalent to the exponent's value. Therefore, 3.47×10^{-4} equals .000347.

8388. Answer A. JSGT 1-11 (AC 65-9A)
When working with powers of 10, a positive exponent indicates you must move the decimal to the right. The number of places the decimal should be moved is equivalent to the exponent's value. Therefore, 16,300 is equal to 1.63×10^4.

8389. Answer C. JSGT 1-11 (AC 65-9A)
Here you must find the square root expressed in scientific notation. The square root of $124.9924 = 11.18$. When working with scientific notation, a negative exponent indicates you must move the decimal to the left; whereas, a positive exponent indicates you should move the decimal to the right. The number of places the decimal is moved is equivalent to the exponent's value. Therefore, 11.18 is equal to $1,118 \times 10^{-2}$.

8390. Answer C. JSGT 1-11 (AC 65-9A)
The square root of 16 is four. Multiply four by itself four times to get an answer of 256.

$$(\sqrt{16})^4 = (4)^4 = 256$$

8391. H01
(Refer to figure 53) Solve the equation.

A — .0297.
B — .1680.
C — .0419.

8391. Answer C. JSGT 1-11 (AC 65-9A)
Begin by calculating the square roots in the numerator and squaring the denominator. Next, add the two values in the numerator and divide the sum by the denominator. The answer is .0419.

$$\frac{\sqrt{31} + \sqrt{43}}{(17)^2} = \frac{5.57 + 6.56}{289} = \frac{12.13}{289} = .0419$$

$$\frac{\sqrt[2]{31} + \sqrt[2]{43}}{(17)^2} =$$

FIGURE 53.—Equation.

8392. H01
The result of 7 raised to the third power plus the square root of 39 is equal to

A — 349.24.
B — .34924.
C — 343.24.

8392. Answer A. JSGT 1-11 (AC 65-9A)
7^3 is equal to 343 ($7 \times 7 \times 7 = 343$). The square root of 39 equals 6.24. The sum of these two values is 349.24.

$$7^3 + \sqrt{39} = 343 + 6.24 = 349.24$$

8393. H01
Find the square root of 1,824.

A — 42.708 × 10 to the negative second power.
B — .42708.
C — .42708 × 10 to the second power.

8393. Answer C. JSGT 1-11 (AC 65-9A)
In this problem you must find the square root expressed in scientific notation. The square root of 1,824 is 42.708. When working with powers of 10, a negative exponent indicates you must move the decimal to the left; whereas, a positive exponent indicates you should move the decimal to the right. The number of places the decimal is moved is equivalent to the exponent's value. Using scientific notation, the answer is $.42708 \times 10^2$.

8409. H03
Select the fraction which is equal to .020.

A — 1/50
B — 1/5
C — 2/5

8409. Answer A. JSGT 1-9 (AC 65-9A)
The decimal value .020 also can be written as

$$\frac{2}{100}.$$

This fraction can be reduced to

$$\frac{2}{100} \div \frac{2}{2} = \frac{1}{50}.$$

8410. H03

1.21875 is equal to

A — 83/64.
B — 19/16.
C — 39/32.

8411. H03

If the volume of a cylinder with the piston at bottom center is 84 cubic inches and the piston displacement is 70 cubic inches, then the compression ratio is

A — 7 to 1.
B — 1.2 to 1.
C — 6 to 1.

8412. H03

Express 7/8 as a percent.

A — 8.75 percent.
B — .875 percent.
C — 87.5 percent.

8413. H03

What is the speed of a spur gear with 42 teeth driven by a pinion gear with 14 teeth turning 420 RPM?

A — 588 RPM.
B — 160 RPM.
C — 140 RPM.

8414. H03

An engine develops 108 horsepower at 87 percent power. What horsepower would be developed at 65 percent power?

A — 80.
B — 70.
C — 64.

8410. Answer C. JSGT 1-9 (AC 65-9A)
To begin, convert the decimal value to a fraction. The fractional equivalent of 1.21875 is

$$1 \frac{21,875}{100,000}$$

This fraction can be reduced to 7/32.

$$\frac{21,875}{100,000} \div \frac{25}{25} = \frac{875}{4,000} \div \frac{25}{25} = \frac{35}{160} \div \frac{5}{5} = \frac{7}{32}.$$

Now convert 1-7/32 to an improper fraction by multiplying the whole number by the denominator and adding it to the numerator. The answer is 39/32.

8411. Answer C. JSGT 1-10 (AC 65-9A)
Compression ratio is the ratio of the cylinder volume with the piston at the bottom of the stroke to the cylinder volume with the piston at the top of the stroke. In this example, the volume with the piston at the bottom of the stroke is 84 cubic inches. To determine the volume with the piston at the top of the stroke you must subtract the piston displacement from 84 cubic inches. This results in a volume of 14 cubic inches (84 − 70 = 14). The compression is 84:14, or 6:1 when simplified.

8412. Answer C. JSGT 1-9 (AC 65-9A)
When converting a fraction to a percentage, divide the numerator by the denominator and multiply the result by 100. The equivalent percentage value of 7/8 is 87.5 percent (7 ÷ 8 = .875 × 100 = 87.5).

8413. Answer C. JSGT 1-10 (AC 65-9A)
To begin, determine the ratio of the two gears. The ratio is 42:14, or 3:1 when simplified. However, since the drive gear is smaller than the gear it's turning, the speed of the driven gear (spur gear) is less than the drive gear (pinion gear). To determine the speed of the spur gear divide the pinion gear RPM by 3. The speed of the spur gear is 140 RPM (420 ÷ 3 = 140).

8414. Answer A. JSGT 1-10 (AC 65-9A)
You can solve this problem by calculating the amount of horsepower generated by 1 percent of power and then multiplying that number by 65. To determine the horsepower generated by 1 percent power, divide 108 HP by 87 percent. Approximately 1.24 HP is generated by each 1 percent of power. Therefore, 65 percent power generates 80.69 HP (1.24 × 65 = 80.69). Answer (A) is closest.

Mathematics

8415. H03
A certain aircraft bolt has an overall length of 1-1/2 inches, with a shank length of 1-3/16 inches, and a threaded portion length of 5/8 inch. What is the grip length?

A — .5625 inch.
B — .8750 inch.
C — .3125 inch.

H.8.4.1.6.B.1 H03
Select the fractional equivalent for a 0.0625 inch-thick sheet of aluminum.

A — 1/16.
B — 1/32.
C — 3/64.

8417. H03
Express 5/8 as a percent.

A — .625 percent.
B — 6.25 percent.
C — 62.5 percent.

8418. H03
Select the decimal which is most nearly equal to 77/64.

A — 0.8311.
B — 0.08311.
C — 1.2031.

8419. H03
An airplane flying a distance of 875 miles used 70 gallons of gasoline. How many gallons will it need to travel 3,000 miles?

A — 250.
B — 240.
C — 144.

8420. H03
What is the speed ratio of a gear with 36 teeth meshed to a gear with 20 teeth?

A — 5 to 12.
B — 6.6 to 12.
C — 5 to 9.

8415. Answer A. JSGT 1-9 (AC 65-9A)
A bolt's shank length goes from the bottom of the head to the end of the shank. On the other hand, grip length goes from the bottom of the head to the beginning of the threads. To determine the grip length of the bolt described in the question, you must convert the bolt's shank length to an improper fraction (1-3/16 = 19/16). Next, convert the threaded portion length to 16ths of an inch (5/8 = 10/16). Subtracting the thread length from the shank length provides the grip length of 9/16 (19/16 − 10/16 = 9/16) or .5625 inch.

8416. Answer A. JSGT 1-9 (AC 65-9A)
To convert a decimal to a common fraction, write the decimal as a fraction and reduce it to its lowest terms. For example, .0625 is equivalent to 625 ten-thousandths or

$$\frac{625}{10,000}$$

This reduces to 1/16.

$$\frac{625}{10,000} \div \frac{625}{625} = \frac{1}{16}$$

8417. Answer C. JSGT 1-9 (AC 65-9A)
To convert a fraction to a percent, divide the numerator (5) by the denominator (8) and multiply the product by 100. The equivalent percent value of 5/8 is 62.5 percent (5 ÷ 8 = .625 × 100 = 62.5).

8418. Answer C. JSGT 1-9 (AC 65-9A)
To convert a fraction to a decimal, divide the numerator (77) by the denominator (64). The decimal equivalent of 77/64 is 1.2031 (77 ÷ 64 = 1.2031).

8419. Answer B. JSGT 1-10 (AC 65-9A)
To solve this problem, you must first calculate how many miles the airplane can fly on one gallon of gas. To do this, divide the miles flown (875) by the number of gallons used (70). The airplane can fly 12.5 miles on one gallon of gas. To determine the amount of gas required to fly 3,000 miles, divide 3,000 by 12.5. A total of 240 gallons (3,000 ÷ 12.5 = 240) are required to fly 3,000 miles.

8420. Answer C. JSGT 1-10 (AC 65-9A)
The ratio of the two gears is 20 to 36. This can be reduced to 5 to 9.

8421. H03
A pinion gear with 14 teeth is driving a spur gear with 42 teeth at 140 RPM. Determine the speed of the pinion gear.

A — 588 RPM.
B — 420 RPM.
C — 240 RPM.

8422. H03
The parts department's profit is 12 percent on a new magneto. How much does the magneto cost if the selling price is $145.60?

A — $128.12.
B — $125.60.
C — $130.00.

8423. H03
If an engine is turning 1,965 rpm at 65 percent power, what is its maximum rpm?

A — 2,653.
B — 3,023.
C — 3,242.

8424. H03
An engine of 98 horsepower maximum is running at 75 percent power. What is the horsepower being developed?

A — 87.00.
B — 33.30.
C — 73.50.

8425. H03
A blueprint shows a hole of 0.17187 to be drilled. Which fraction size drill bit is most nearly equal?

A — 11/64.
B — 9/32.
C — 11/32.

8426. H03
Which decimal is most nearly equal to a bend radius of 31/64?

A — 0.2065.
B — 0.4844.
C — 0.3164.

8421. Answer B. JSGT 1-10 (AC 65-9A)
To begin, determine the ratio of the two gears. The ratio is 14:42, or 1:3 when simplified. This means that for every one turn of the driven gear (spur gear), the drive gear (pinion gear) turns three times. Therefore, if the driven gear turns at 140 RPM, the drive gear will turn at 420 RPM (140 × 3 = 420).

8422. Answer C. JSGT 1-9 (AC 65-9A)
If the price includes a 12 percent profit, then $145.60 is equal to 112 percent of the cost. To determine the magneto's cost, divide $145.60 by 112 percent to find what 1 percent is worth. In this case, 1 percent of the price is $1.30 (145.60 ÷ 112 = 1.30). Now, multiply the 1 percent price by 100 percent to get the magneto's cost of $130.00 (1.30 × 100 = 130.00). Answer (A) is incorrect because $128.12 represents the price if 12 percent of the selling price is deducted from the selling price.

8423. Answer B. JSGT 1-10 (AC 65-9A)
To determine the maximum rpm, you must divide the known rpm by the decimal equivalent of the percentage. Based on this, the engine turns at a maximum of 3,023 rpm (1,965 ÷ .65 = 3,023).

8424. Answer C. JSGT 1-9 (AC 65-9A)
If an engine has a maximum horsepower of 98 HP and the engine is run at 75 percent power, then the amount of horsepower developed equals 75 percent times 98 HP. The amount of horsepower developed is 73.5 HP (98 × .75 = 73.5).

8425. Answer A. JSGT 1-9 (AC 65-9A)
To convert 0.17187 to a common fraction, rewrite it as a fraction and reduce it to its simplest terms

$$\frac{17,187}{100,000} = \frac{11}{64}$$

Another way to solve this problem is to divide the numerator by the denominator for each of the three choices and find which is closest to 0.17187.

8426. Answer B. JSGT 1-9 (AC 65-9A)
To convert a fraction to a decimal, divide the numerator by the denominator. The decimal equivalent of 31/64 is .484375 (31 ÷ 64 = .484375). Answer (B) is the closest.

8427.   **H03**
Sixty-five engines are what percent of 80 engines?

A — 81 percent.
B — 65 percent.
C — 52 percent.

8428.   **H03**
The radius of a piece of round stock is 7/32. Select the decimal which is most nearly equal to the diameter.

A — 0.2187.
B — 0.4375.
C — 0.3531.

8429.   **H03**
Maximum engine life is 900 hours. Recently, 27 engines were removed with an average life of 635.3 hours. What percent of the maximum engine life has been achieved?

A — 71 percent.
B — 72 percent.
C — 73 percent.

8430.   **H03**
What is the ratio of 10 feet to 30 inches?

A — 4:1.
B — 1:3.
C — 3:1.

8435.   **H04**
What is the ratio of a gasoline fuel load of 200 gallons to one of 1,680 pounds?

A — 5:7.
B — 2:3.
C — 5:42.

8438.   **H04**
(Refer to figure 59) Solve the equation.

A — +31.25.
B — −5.20.
C — −31.25.

8427. Answer A. JSGT 1-9 (AC 65-9A)
This problem asks what percentage 65 is of 80. To calculate this, divide 65 by 80 and multiply the answer by 100. The answer is 81.25 percent (65 ÷ 80 = .8125 × 100 = 81.25). Answer (A) is the closest.

8428. Answer B. JSGT 1-9 (AC 65-9A)
To convert a fraction to a decimal, divide the numerator by the denominator. The decimal equivalent of 7/32 is .21875 (7 ÷ 32 = .21875). To determine the diameter, multiply the radius by two. The diameter is 0.4375 (.21875 × 2 = .4375).

8429. Answer A. JSGT 1-9 (AC 65-9A)
The question asks what percent 635.3 is of 900. To calculate this, divide 635.3 by 900 and multiply the answer by 100. The answer is 70.59 (635.3 ÷ 900 = .7059 × 100 = 70.59). Answer (A) is the closest.

8430. Answer A. JSGT 1-10 (AC 65-9A)
To get a ratio, both measurements must be in inches. Ten feet is equal to 120 inches. The ratio is 120:30 which reduces to 4:1.

8435. Answer A. JSGT 1-10 (AC 65-9A)
In a ratio, both numbers must be in like terms. Because of this, you must convert gallons to pounds. One gallon of gasoline weighs 6 lbs., so 200 gallons weighs 1,200 lbs. The ratio now becomes 1,200:1,680. This simplifies to 5:7.

8438. Answer B. JSGT 1-4 (AC 65-9A)
This problem involves the multiplication and division of signed numbers. Remember, division or multiplication of unlike signs always results in a negative number, whereas division or multiplication of like signs results in a positive number. The answer is −5.20.

$$\frac{-4\sqrt{125}}{-6\sqrt{-36}} =$$

FIGURE 59.—Equation.

SECTION B
ALGEBRA

This section introduces you to the basic operations of algebra. It covers solving for a variable, the correct order of operations, and solving complex equations. FAA Test questions that apply to this section include:

8381, 8432, 8433, 8434, 8436, 8437, 8439, 8440, 8441, 8442.

8381.   **H01**
(Refer to figure 52) Solve the equation.

A — 115.
B — 4.472
C — 5.

8381. Answer C. JSGT 1-14 (AC 65-9A)
To solve this problem, begin by solving everything in parentheses. Remember, any number raised to the zero power is 1. Next, multiply where appropriate and then add. The answer is 5.

$$\sqrt{(-4)^0 + 6 + (\sqrt[4]{1296})(\sqrt{3})^2} = \sqrt{[1 + 6 + (6 \times 3)]}$$
$$= \sqrt{(1 + 6 + 18)}$$
$$= \sqrt{25}$$
$$= 5$$

$$\sqrt{(-4)^0 + 6 + (\sqrt[4]{1296})(\sqrt{3})^2} =$$

FIGURE 52.—Equation.

8432.   **H04**
Solve the equation.
$[(4 \times -3) + (-9 \times 2)] \div 2 =$

A — -30.
B — -15.
C — -5.

8432. Answer B. JSGT 1-15 (AC 65-9A)

Complex problems such as this require the operations to be done in a particular order. Begin by doing everything in parentheses. This simplifies the equation to $[-12 + -18] \div 2$. Now, solve what is in the brackets and divide the result by 2. The answer is -15.

$$[-12 + -18] \div 2 = -30 \div 2 = -15.$$

8433.   **H04**
Solve the equation.
$(64 \times 3/8) \div 3/4 =$

A — 16.
B — 24.
C — 32.

8433. Answer C. JSGT 1-15 (AC 65-9A)
To begin, convert the fractions in the equation to decimal numbers by dividing the numerator by the denominator. The decimal equivalent to 3/8 is .375 ($3 \div 8 = .375$) and to 3/4 is .75 ($3 \div 4 = .75$). When solving any equation, you must do what is in parentheses first. Once this is done, you can divide by .75. The answer is 32.

8434. H04
Solve the equation.
$(32 \times 3/8) \div 1/6 =$

A — 12.
B — 2.
C — 72.

8436. H04
Solve the equation.
$2/4 (30 + 34) 5 =$

A — 160.
B — 245.
C — 640.

8437. H04
(Refer to figure 58) Solve the equation.

A — 174.85.
B — −81.49.
C — 14.00.

8434. Answer C. JSGT 1-14 (AC 65-9A)
To begin, convert the fractions in the equation to decimal numbers by dividing the numerator by the denominator. The decimal equivalent of 3/8 is .375 (3 ÷ 3 = .375) and to 1/6 is .167 (1 ÷ 6 = .167). When solving any equation, you must do what is in parentheses first. Once this is done, you can divide by .167. The answer is 71.86.

$$(32 \times .375) \div .167 = 12 \div .167 = 71.86.$$

Answer (C) is closest.

8436. Answer A. JSGT 1-14 (AC 65-9A)
To begin, convert the fraction in the equation to a decimal number by dividing the numerator by the denominator. The decimal equivalent to 2/4 is .5 (2 ÷ 4 = .5). When solving any equation, you must do what is in parentheses first. Once this is done, you can multiply from left to right. The answer is 160.

$$.5 (30 + 34) 5 = .5 (64) 5 = 160.$$

8437. Answer C. JSGT 1-15 (AC 65-9A)
When solving complex equations, begin by solving everything in parentheses first. Once this is done you can perform multiplication followed by addition. The answer is 14.00.

$$\frac{(-35 + 25)(-7) + (\pi)(16^{-2})}{\sqrt{25}} = \frac{(-10)(-7) + (3.1416)(.0039)}{5}$$

$$= \frac{70 + .0123}{5}$$

$$= \frac{70.0123}{5}$$

$$= 14.00$$

$$\frac{(-35 + 25)(-7) + (\pi)(16^{-2})}{\sqrt{25}} =$$

FIGURE 58.—Equation.

8439.   **H04**
Solve the equation.
$4 - 3[-6(2 + 3) + 4] =$

A — 82.
B — –25.
C — –71.

8439. Answer A. JSGT 1-14 (AC 65-9A)
When solving any equation, you must do what is in parentheses first, then brackets. This is followed by the operation of multiplication and then addition. The answer is 82.

$$4 - 3[-6(2+3) + 4] = 4 - 3[-6 \times 5 + 4]$$
$$= 4 - 3[-26]$$
$$= 4 - (-78)$$
$$= 82$$

8440.   **H04**
Solve the equation.
$-6[-9(-8+4) - 2(7+3)] =$

A — –332.
B — 216.
C — –96.

8440. Answer C. JSGT 1-14 (AC 65-9A)
When solving any equation, you must do what is in parentheses first, then brackets. This is followed by the operation of multiplication and then addition. The answer is –96.

$$-6[-9(-8+4) - 2(7+3)] = -6[-9(-4) - 2(10)]$$
$$= -6[36 - 20]$$
$$= -6[16]$$
$$= -96$$

8441.   **H04**
Solve the equation.
$(-3 + 2)(-12 - 4) + (-4 + 6) \times 2$

A — 20.
B — 35.
C — 28.

8441. Answer A. JSGT 1-14 (AC 65-9A)
When solving any equation, you must do what is in parentheses first. This is followed by the operation of multiplication and then addition. The answer is 20.

$$(-3+2)(-12-4) + (-4+6) \times 2 = (-1)(-16) + (2) \times 2$$
$$= 16 + 4$$
$$= 20$$

8442.   **H04**
(Refer to figure 60) Solve the equation.

A — 11.9.
B — 11.7.
C — 11.09.

8442. Answer A. JSGT 1-14 (AC 65-9A)
When solving complex equations, begin by solving everything in parentheses first. Once this is done you can perform multiplication and division followed by addition. The answer is 11.9.

$$\frac{(-5+23)(-2) + (3^{-3})(\sqrt{64})}{-27 \div 9} = \frac{18(-2) + (.037)(8)}{-3}$$
$$= \frac{-36 + .2963}{-3}$$
$$= \frac{-35.7037}{-3}$$
$$= 11.9$$

$$\frac{(-5 + 23)(-2) + (3^{-3})(\sqrt{64})}{-27 \div 9} =$$

FIGURE 60.—Equation.

Mathematics

SECTION C
GEOMETRY AND TRIGONOMETRY

Section C of Chapter 1 applies the math techniques you learned in the previous two sections. It discusses the computation of area, trigonometric functions, the metric system and metric conversions, and the use of math hardware such as electronic calculators. FAA Test questions drawn from this section include:

8394, 8395, 8396, 8397, 8399, 8400, 8401, 8402, 8403, 8404, 8405, 8406, 8407, 8408.

8394.   H02
The total piston displacement of a specific engine is

A — dependent on the compression ratio.
B — the volume displaced by all the pistons during one revolution of the crankshaft.
C — the total volume of all the cylinders.

8394. Answer B. JSGT 1-20 (AC 65-9A)
Piston displacement is the volume of air displaced by a piston as it moves from bottom dead center to top dead center. Total piston displacement of an engine equals the volume displaced by all the pistons during one revolution of the crankshaft. The compression ratio (answer A) is a comparison of the volume of space in a cylinder when the piston is at the bottom of the stroke, to the volume of space when the piston is at the top of the stroke, and is, therefore, incorrect. Answer (C) is incorrect because the total volume of all cylinders does not deduct the volume of the pistons.

8395.   H02
(Refer to figure 54) Compute the area of the trapezoid.

A — 52.5 square feet.
B — 60 square feet.
C — 76.5 square feet.

8395. Answer A. JSGT 1-18 (AC 65-9A)
To compute the area of a trapezoid, use the formula

$$A = \frac{1}{2}(b_1 + b_2)h$$

The answer is 52.5 square feet. A = 1/2 (9 + 12) 5 = 52.5 square feet.

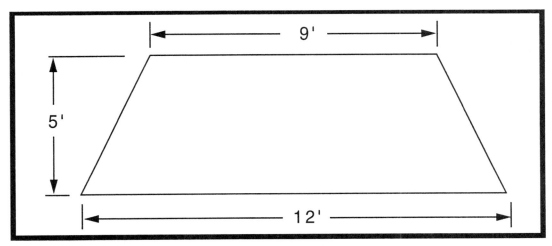

FIGURE 54.—Trapezoid Area.

8396.   H02
What size sheet of metal is required to fabricate a cylinder 20 inches long and 8 inches in diameter? (Note: $C = \pi \times D$)

A — 20" × 25-5/32".
B — 20" × 24-9/64".
C — 20" × 25-9/64".

8396. Answer C. JSGT 1-18 (AC 65-9A)
The height of the cylinder is 20 inches. Therefore, the height of the sheet of metal must be 20 inches. To determine the length required, you must calculate the circumference of the cylinder using the formula given. The circumference is 25.132 ($C = \pi \times D = 3.1416 \times 8 = 25.132$) which is slightly smaller than 25-9/64. Therefore, a 20 inch × 25-9/64 inch sheet of metal is required to fabricate the cylinder.

8397. **H02**
(Refer to figure 55) Find the area of the right triangle shown.

A — 12 square inches.
B — 6 square inches.
C — 15 square inches.

8397. Answer B. JSGT 1-17 (AC 65-9A)
The area of a triangle is calculated using the formula:

$$A = 1/2\ bh.$$

The base (b) of the triangle is 4 inches, and the height (h) is 3 inches. The area of the triangle is 6 square inches ($1/2 \times 4 \times 3 = 6$).

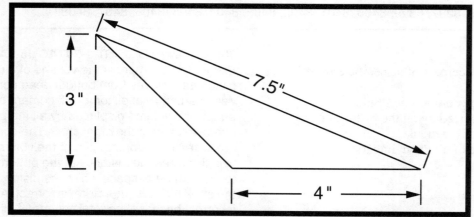

FIGURE 55.—Triangle Area.

8399. **H02**
A rectanglar-shaped fuel tank measures 60 inches in length, 30 inches in width, and 12 inches in depth. How many cubic feet are within the tank?

A — 12.5
B — 15.0
C — 21.0

8399. Answer A. JSGT 1-19 (AC 65-9A)
To determine the volume of a rectangle, multiply the length (L) times the width (W) times the depth (D). However, the question asks for cubic feet and the dimensions are given in inches. Therefore, you must convert the inches into feet and then compute the volume. The answer is 12.5 cubic feet (5 feet × 2.5 feet × 1 foot = 12.5 cubic feet).

8400. **H02**
Select the container size that will be equal in volume to 60 gallons of fuel.
(7.5 gal = 1 cu. ft.)

A — 7.5 cubic feet.
B — 8.0 cubic feet.
C — 8.5 cubic feet.

8400. Answer B. JSGT 1-18 (AC 65-9A)
Using the given relationship of 7.5 gal = 1 cu. ft., you can determine the number of cubic feet required to hold 60 gallons by dividing 60 by 7.5. The answer is 8 cubic feet (60 ÷ 7.5 = 8).

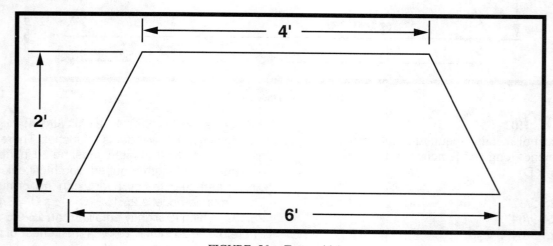

FIGURE 56.—Trapezoid Area.

8401. **H02**
(Refer to figure 56) Compute the area of the trapezoid.

A — 24 square feet.
B — 48 square feet.
C — 10 square feet.

8402. **H02**
(Refer to figure 57) Determine the area of the triangle formed by points A, B, and C.
A to B = 7.5 inches.
A to D = 16.8 inches.

A — 42 square inches.
B — 63 square inches.
C — 126 square inches.

8401. Answer C. JSGT 1-18 (AC 65-9A)
To compute the area of a trapezoid use the formula

$$A = \frac{1}{2}(b_1 + b_2)h$$

The answer is 10 square feet. A = 1/2 (4 + 6) 2 = 10 square feet.

8402. Answer B. JSGT 1-17 (AC 65-9A)
The area of a triangle is calculated using the formula

$$A = \frac{1}{2}bh.$$

The base (b) in this case is the distance from B to C, which is the same as the distance from A to D or 16.8 inches. The height (h) is the distance from A to B or 7.5 inches. Therefore, 1/2 × (16.8 × 7.5) = 63.

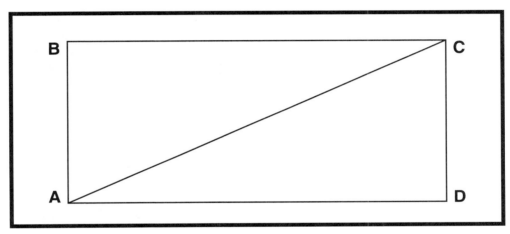

FIGURE 57.—Triangle Area.

8403. **H02**
What is the piston displacement of a master cylinder with a 1.5-inch diameter bore and a piston stroke of 4 inches?

A — 9.4247 cubic inches.
B — 7.0686 cubic inches.
C — 6.1541 cubic inches.

8403. Answer B. JSGT 1-20 (AC 65-9A)
When asked to compute piston displacement, you must calculate the volume of the cylinder. To calculate the volume of a cylinder, use the formula:

$$V = \pi r^2 h$$

Where (V) is the volume, (r) is the radius of the cylinder, and (h) is the height or stroke of the piston. The displacement is 7.0686 cubic inches

$$(3.1416 \times .75^2 \times 4) = 7.0686.$$

8404. **H02**
How many gallons of fuel will be contained in a rectangular-shaped tank which measures 2 feet in width, 3 feet in length, and 1 foot 8 inches in depth?
(7.5 gal = 1 cu. ft.)

A — 66.6.
B — 75.
C — 45.

8404. Answer B. JSGT 1-19 (AC 65-9A)
The first step in solving this problem is to calculate the volume of the tank. This is done by multiplying the length times the width times the height. The volume is 10 cubic feet (3 × 2 × 1.667 = 10). Using the given relationship of 7.5 gal. = 1 cu. ft., you can determine the number of gallons the tank will hold by multiplying the volume times 7.5 gallons. The tank holds 75 gallons (10 × 7.5 = 75).

8405.　　　**H02**
A rectangular-shaped fuel tank measures 27-1/2 inches in length, 3/4 foot in width, and 8-1/4 inches in depth. How many gallons will the tank contain?

(231 cu. in. = 1 gal.)

A — 7.86.
B — 8.80.
C — 9.80.

8406.　　　**H02**
A four-cylinder aircraft engine has a cylinder bore of 3.78 inches and is 8.5 inches deep. With the piston on bottom center, the top of the piston measures 4.0 inches from the bottom of the cylinder. What is the approximate piston displacement of this engine?

A — 200 cubic inches.
B — 360 cubic inches.
C — 235 cubic inches.

8407.　　　**H02**
A rectangular-shaped fuel tank measures 37-1/2 inches in length, 14 inches in width, and 8-1/4 inches in depth. How many cubic inches are within the tank?

A — 525.
B — 433.125.
C — 4,331.25.

8408.　　　**H02**
A six-cylinder engine with a bore of 3.5 inches, a cylinder height of 7 inches and a stroke of 4.5 inches will have a total piston displacement of

A — 256.88 cubic inches.
B — 259.77 cubic inches.
C — 43.3 cubic inches.

8405. Answer B. JSGT 1-19 (AC 65-9A)
The first step in solving this problem is to calculate the volume of the tank. This is done by converting all dimensions to inches and then multiplying the length times the width times the height. The volume is 2,041.875 cubic inches ($27.5 \times 9 \times 8.25 = 2,041.875$). Using the given relationship of 231 cubic inches = 1 gallon, you can determine the number of gallons the tank will hold by dividing the volume by 231 cubic inches. The tank holds 8.84 gallons ($2,041.875 \div 231 = 8.84$). Of the choices given, answer (B) is the closest.

8406. Answer A. JSGT 1-20 (AC 65-9A)
When asked to compute displacement, you must calculate the volume of cylinder. To calculate the volume of a cylinder, use the formula:

$$V = \pi r^2 h,$$

Where (v) is the volume, (r) is the radius of the cylinder, and (h) is the height or stroke of the piston measured from the top of the piston to the top of the cylinder. The displacement of each cylinder is 50.5 cubic inches ($3.1416 \times 1.89^2 \times 4.5 = 50.5$). To determine the displacement of the entire engine, multiply the displacement of each cylinder by the number of cylinders. The engine displacement is 202 cubic inches ($50.5 \times 4 = 202$). Answer (A) is the closest.

8407. Answer C. JSGT 1-19 (AC 65-9A)
When calculating the volume of an object multiply the length times the width times the height. The volume is 4,331.25 cubic inches ($37.5 \times 14 \times 8.25 = 4,331.25$).

8408. Answer B. JSGT 1-20 (AC 65-9A)
When asked to compute displacement, you must calculate the volume of the cylinder. To calculate the volume of a cylinder, use the formula:

$$V = \pi r^2 h$$

Where (V) is the volume, (r) is the radius of the cylinder, and (h) is the height or stroke of the piston. The displacement of each cylinder is 43.295 cubic inches ($3.1416 \times 1.75^2 \times 4.5 = 43.295$). To determine the displacement of the entire engine, multiply the displacement of each cylinder by the number of cylinders. The engine displacement is 259.77 ($43.295 \times 6 = 259.77$).

CHAPTER 2

PHYSICS

SECTION A
MATTER AND ENERGY

The first secton of Chapter 2 introduces some basic physical concepts, including the chemical nature of matter, the physical nature of matter, and energy. Although there are is only one FAA Test question in this area, much of the information presented provides a base for items presented in future sections.

8466.

8466.   J01
The boiling point of a given liquid varies

A — directly with pressure.
B — inversely with pressure.
C — directly with volume.

8466. Answer A. JSGT 2-4 (AC 65-9A)
If two values are directly related, increasing one value will increase the other value. The boiling point of a liquid goes up as the pressure goes up, so they are said to be directly related.

SECTION B
WORK, POWER, FORCE, AND MOTION

Section B of Chapter 2 contains information on power, force, stress, strain, and motion. This section includes information for the following FAA Test questions:

8468, 8471, 8477, 8480, 8481.

8468.   J01
An engine that weighs 350 pounds is removed from an aircraft by means of a mobile hoist. The engine is raised 3 feet above its attachment mount, and the entire assembly is then moved forward 12 feet. A constant force of 70 pounds is required to move the loaded hoist. What is the total work input required to move the hoist?

A — 840 foot-pounds.
B — 1,890 foot-pounds.
C — 1,050 foot-pounds.

8468. Answer A. JSGT 2-7 (AC 65-9A)
Work is determined by the formula $W = F \times D$, where (F) is the force applied and (D) represents the distance moved. The question asks for the work required to move the hoist only; therefore, the work required is 840 foot-pounds (70 pounds × 12 feet = 840 foot-pounds). Answer (C), 1,050 foot-pounds represents the work required to lift the engine and answer (B), 1,890 foot-pounds is the amount of work required to lift the engine and move the hoist.

8471.　　　　J01
(Refer to figure 61) The amount of force applied to rope A to lift the weight is

A — 12 pounds.
B — 15 pounds.
C — 20 pounds.

8471. Answer B. JSGT 2-12 (AC 65-9A)
The mechanical advantage of a pulley system is equal to the number of ropes supporting the resistance minus the rope you are pulling on. In this example, the mechanical advantage is four. This means that for every one pound of effort exerted on the rope, four pounds are lifted. In this problem, 60 pounds requires a force of 15 pounds (60 ÷ 4 = 15).

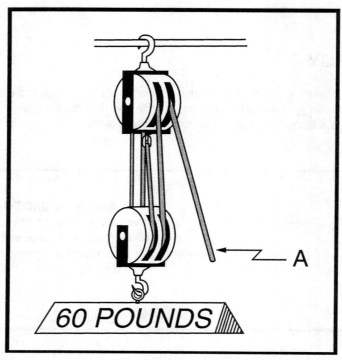

FIGURE 61.—Physics.

8477.　　　　J01
How much work input is required to lower (not drop) a 120-pound weight from the top of a 3-foot table to the floor?

A — 120 pounds of force.
B — 360 foot-pounds.
C — 40 foot-pounds.

8477. Answer B. JSGT 2-7 (AC 65-9A)
Work is calculated through the formula W = F × D, where (F) represents the applied force or weight and (D) represents the distance the object is moved. In this example, 360 foot-pounds of work are required (120 × 3 = 360).

8480.　　　　J01
How many, if any, factors are necessary to determine power?

　　1. Force exerted.
　　2. Distance the force moves.
　　3. Time required to do the work.

A — One.
B — Two.
C — Three.

8480. Answer C. JSGT 2-7 (AC 65-9A)
To determine power (P), you must know the force (F) that is exerted, the distance (D) the force moves the object, and the time (t) required to do the work. This can be seen in the formula

$$P = \frac{(F \times D)}{t}$$

Physics

8481.   J01
What force must be applied to roll a 120-pound barrel up an inclined plane 9 feet long to a height of 3 feet (disregard friction)?

$$L \div I = R \div E$$

L = Length of ramp, measured along the slope
I = Height of ramp
R = Weight of object to be raised or lowered
E = Force required to raise or lower object

A — 40 pounds.
B — 120 pounds.
C — 360 pounds.

8481. Answer A. JSGT 2-10 (AC 65-9A)
To determine the force, or effort, required to move the barrel up the ramp, use the formula given (L ÷ I = R ÷ E). The required force is 40 pounds.

$$9 \div 3 = 120 \div E$$
$$3 = \frac{120}{E}$$
$$3E = 120$$
$$E = 40$$

SECTION C
GAS AND FLUID MECHANICS

Section C discusses basic gas and fluid mechanics. Included in this section is information on temperature, pressure, the gas laws, fluid mechanics, and sound. The following FAA Test questions apply to this section:

8216, 8398, 8465, 8467, 8470, 8475, 8476, 8479, 8482, 8485.

8216.   D01
Which statement concerning Bernoulli's principle is true?

A — The pressure of a fluid increases at points where the velocity of the fluid increases.
B — The pressure of a fluid decreases at points where the velocity of the fluid increases.
C — It applies only to gases.

8216. Answer B. JSGT 2-28 (AC 65-9A)
Bernoulli's Principle states that when a fluid is in motion, the outward pressure exerted decreases as the fluid velocity increases.

8398.   H02
What force is exerted on the piston in a hydraulic cylinder if the area of the piston is 1.2 square inches and the fluid pressure is 850 PSI?

A — 1,020 pounds.
B — 960 pounds.
C — 850 pounds.

8398. Answer A. JSGT 2-28 (AC 65-9A)
Pascal's Law states that any force applied to a confined fluid is transmitted equally in all directions. The amount of force (F) applied can be determined by multiplying the fluid pressure (P) times the area (A) in which the fluid is contained. The force exerted on the piston is 1,020 pounds (850 × 1.2 = 1,020).

8465.   J01
The force that can be produced by an actuating cylinder whose piston has a cross-sectional area of 3 square inches operating in a 1,000 PSI hydraulic system is most nearly

A — 3,000 pounds.
B — 334 pounds.
C — 1,000 pounds.

8465. Answer A. JSGT 2-28 (AC 65-9A)
According to Pascal's Law, Force = Pressure × Area. Substituting the values given, the force produced is 3,000 pounds (1,000 PSI × 3 sq. in. = 3,000 pounds).

8467. J01
Which of the following is NOT considered a method of heat transfer?

A — Convection.
B — Conduction.
C — Diffusion.

8470. J01
Under which conditions will the rate of flow of a liquid through a metering orifice (or jet) be the greatest (all other factors being equal)?

A — Unmetered pressure — 18 PSI,
    metered pressure — 17.5 PSI,
    atmospheric pressure — 14.5 PSI.
B — Unmetered pressure — 23 PSI,
    metered pressure — 12 PSI,
    atmospheric pressure — 14.3 PSI.
C — Unmetered pressure — 17 PSI,
    metered pressure — 5 PSI,
    atmospheric pressure — 14.7 PSI.

8475. J01
If the volume of a confined gas is doubled (without the addition of more gas), the pressure will (assume the temperature remains constant)

A — increase in direct proportion to the volume increase.
B — remain the same.
C — be reduced to one-half its original value.

8476. J01
If the temperature of a confined liquid is held constant and its pressure is tripled, the volume will

A — triple.
B — be reduced to one-third its original volume.
C — remain the same.

8479. J01
If the fluid pressure is 800 PSI in a 1/2-inch line supplying an actuating cylinder with a piston area of 10 square inches, the force exerted on the piston will be

A — 4,000 pounds.
B — 8,000 pounds.
C — 800 pounds.

8482. J01
Which statement concerning heat and/or temperature is true?

A — There is an inverse relationship between temperature and heat.
B — Temperature is a measure of the kinetic energy of the molecules of any substance.
C — Temperature is a measure of the potential energy of the molecules of any substance.

8467. Answer C. JSGT 2-19 (AC 65-9A)
Heat is transferred by three methods: conduction, convection, and radiation.

8470. Answer C. JSGT 2-28 (AC 65-9A)
Bernoulli's principle states that the pressure of a fluid decreases at points where the velocity of the fluid increases. Therefore, the flow rate (velocity) in this problem is greatest when the difference between the metered and unmetered pressure is greatest (answer C).

8475. Answer C. JSGT 2-23 (AC 65-9A)
Boyle's law states that the volume of an enclosed dry gas varies inversely with its pressure, provided the temperature remains constant. Therefore, when the volume of a gas is doubled, the pressure is cut in half.

8476. Answer C. JSGT 2-23 (AC 65-9A)
For all practical purposes, liquids are considered incompressible. In other words, an increase in fluid pressure is accompanied by a negligible reduction in volume.

8479. Answer B. JSGT 2-28 (AC 65-9A)
The size of the supply line has no affect on the answer to this problem. Simply apply Pascal's Law (Force = Pressure × Area). The force exerted on the piston is 8,000 pounds (800 PSI × 10 sq. in. = 8,000 pounds).

8482. Answer B. JSGT 2-21 (AC 65-9A)
Heat is a form of energy that causes molecular agitation within a material. The amount of agitation is measured by temperature (answer B). The temperature at which molecular motion stops is known as Absolute Zero.

Physics

8485.   J01
If both the volume and the absolute temperature of a confined gas are doubled, the pressure will

A — not change.
B — be halved.
C — become four times as great.

8485. Answer A. JSGT 2-24 (AC 65-9A)
The relationship between volume, temperature, and pressure of a confined gas is described by the General Gas Law. If both the volume and the absolute temperature are doubled, their effects cancel out and the pressure remains constant.

SECTION D
AERODYNAMICS

Maintenance technicians must understand the atmospheric forces that act on aircraft. Chapter 2, Section D discusses the physics of flight and aerodynamics, as well as the forces of lift and drag. It introduces the axes of an airplane, stability, and aircraft flight controls and trim systems, including those of large aircraft. FAA Test questions relevant to this section include:

   8469, 8472, 8473, 8478, 8483, 8484, 8486, 8487, 8488, 8489, 8490, 8491.

8469.   J01
Which condition is the actual amount of water vapor in a mixture of air and water?

A — Relative humidity.
B — Dewpoint.
C — Absolute humidity.

8469. Answer C. JSGT 2-34 (AC 65-9A)
The actual amount of water vapor in a mixture of air and water is known as absolute humidity. Relative humidity (answer A) is the ratio of the amount of water vapor present in the atmosphere to the amount that would be present if the air were saturated. Dewpoint (answer B) is the temperature to which the air must be cooled to become saturated.

8472.   J01
Which will weigh the least?

A — 98 parts of dry air and 2 parts of water vapor.
B — 35 parts of dry air and 65 parts of water vapor.
C — 50 parts of dry air and 50 parts of water vapor.

8472. Answer B. JSGT 2-33 (AC 65-9A)
Humid air at a given temperature and pressure is lighter than dry air at the same temperature and pressure. Therefore, the choice with the greatest proportion of water vapor (answer B) weighs the least.

8473.   J01
Which is the ratio of the water vapor actually present in the atmosphere to the amount that would be present if the air were saturated at the prevailing temperature and pressure?

A — Absolute humidity.
B — Relative humidity.
C — Dewpoint.

8473. Answer B. JSGT 2-34 (AC 65-9A)
Relative humidity is the ratio of the amount of water vapor actually present in the atmosphere to the amount that would be present if the air were saturated at the prevailing temperature and pressure. Absolute humidity (answer A) is the actual amount of water vapor in a mixture of air and water, and dewpoint (answer C) is the temperature to which the air must be cooled to become saturated.

8478.   J01
Which atmospheric conditions will cause the true landing speed of an aircraft to be the greatest?

A — Low temperature with low humidity.
B — High temperature with low humidity.
C — High temperature with high humidity.

8478. Answer C. JSGT 2-33 (AC 65-9A)
True airspeed (TAS) represents the true speed of an airplane through the air. As air temperature and humidity increase, the density of the air decreases. As air density decreases, true airspeed increases. Therefore, high temperature with high humidity will cause an aircraft's landing speed to be greatest.

**8483.	J01**
What is absolute humidity?

A — The temperature to which humid air must be cooled at constant pressure to become saturated.
B — The actual amount of the water vapor in a mixture of air and water.
C — The ratio of the water vapor actually present in the atmosphere to the amount that would be present if the air were saturated at the prevailing temperature and pressure.

**8484.	J01**
The temperature to which humid air must be cooled at constant pressure to become saturated is called

A — dewpoint.
B — absolute humidity.
C — relative humidity.

**8486.	J01**
If all, or a significant part of a stall strip is missing on an airplane wing, a likely result will be

A — increased lift in the area of installation on the opposite wing at high angles of attack.
B — asymmetrical aileron control at low angles of attack.
C — asymmetrical aileron control at or near stall angles of attack.

**8487.	J01**
An airplane wing is designed to produce lift resulting from relatively

A — positive air pressure below and above the wing's surface.
B — negative air pressure below the wing's surface and positive air pressure above the wing's surface.
C — positive air pressure below the wing's surface and negative air pressure above the wing's surface.

**8488.	J01**
The purpose of aircraft wing dihedral is to

A — increase lateral stability.
B — increase longitudinal stability.
C — increase lift coefficient of the wing.

**8489.	J01**
Aspect ratio of a wing is defined as the ratio of the

A — wingspan to the wing root.
B — square of the chord to the wingspan.
C — wingspan to the mean chord.

8483. Answer B. JSGT 2-34 (AC 65-9A)
Absolute humidity is the actual amount of water vapor present in a mixture of air and water. It is usually measured in grams per cubic meter or pounds per cubic foot. Dewpoint is the temperature to which humid air must be cooled at a constant pressure to become saturated (answer A), whereas the ratio of the water vapor actually present in the atmosphere to the amount that would be present if the air were saturated at the prevailing temperature and pressure (answer C) is relative humidity.

8484. Answer A. JSGT 2-34 (AC 65-9A)
Dewpoint is the temperature to which humid air must be cooled at a constant pressure to become saturated.

8486. Answer C. JSGT 2-42 (AC 61-21A)
A stall strip is a small wedge attached to a wing's leading edge that causes the inboard portion of the wing to stall before the outboard portion. This allows the ailerons to maintain effectiveness up to the point of full stall. If part of a stall strip is missing, asymmetrical aileron control will result at or near stall angles of attack.

8487. Answer C. JSGT 2-36 (AC 65-9A)
As air flows over an airfoil it creates an area of high (positive) pressure below the wing and an area of low (negative) pressure above the wing. This differential in pressure is responsible for the creation of lift.

8488. Answer A. JSGT 2-51 (AC 61-21A)
Lateral stability, or roll stability, is increased through the use of dihedral. In other words, the wings on either side of the airplane join the fuselage to form a slight V called dihedral.

8489. Answer C. JSGT 2-42 (AC 61-21A)
A wing's aspect ratio is the ratio of the wing span to the average, or mean, chord.

Physics

8490.   **J01**
A wing with a very high aspect ratio (in comparison with a low aspect ratio wing) will have

A — increased drag at high angles of attack.
B — a low stall speed.
C — poor control qualities at low airspeed.

8490. Answer B. JSGT 2-42 (AC 61-21A)
A wing with a high aspect ratio has low wing loading and, therefore, stalls at a lower speed.

8491.   **J01**
An increase in the speed at which an airfoil passes through the air increases lift because

A — the increased speed of the airflow creates a greater pressure differential between the upper and lower surfaces.
B — the increased speed of the airflow creates a lesser pressure differential between the upper and lower surfaces.
C — the increased velocity of the relative wind increases the angle of attack.

8491. Answer A. JSGT 2-40 (AC 65-9A)
The faster an airfoil moves through the air, the greater the pressure differential between the upper and lower surfaces. The greater the pressure differential, the greater the lift.

SECTION E
HIGH-SPEED AERODYNAMICS

Section E describes factors affecting supersonic flight. Information on the compressibility of air, the speed of sound, supersonic flow patterns, and airfoils suitable to high-speed flight are discussed in detail. Critical mach number, supersonic engine inlets, and the problem of aerodynamic heating are described. The FAA Test question covering these areas is:

    8474.

8474.   **J01**
The speed of sound in the atmosphere

A — varies according to the frequency of the sound.
B — changes with a change in temperature.
C — changes with a change in pressure.

8474. Answer B. JSGT 2-52 (AC 65-9A)
The speed of sound in the atmosphere varies with temperature. As the temperature decreases the speed of sound also decreases, and as temperature increases the speed of sound increases.

SECTION F
HELICOPTER AERODYNAMICS

The last section of Chapter 2 discusses rotary-wing aerodynamics. Some of the topics brought out in this section include translational lift, dissymmetry of lift, and gyroscopic precession. There are no FAA Test questions on this material. However, as a maintenance technician, you must be familiar with basic helicopter flight principles.

CHAPTER 3

BASIC ELECTRICITY

SECTION A
THEORY AND PRINCIPLES

The first section of Chapter 3 introduces the basic theory and principles of electricity. The section begins by discussing how electricity was discovered and continues by explaining electron theory, static electricity, magnetism, electromagnetism, and sources of electricity. Additional information contained in this section includes the relationships explained by Ohm's law as well as information on basic circuit elements and circuit considerations. FAA Test questions pertinent to this section include:

8016, 8017, 8019, 8020, 8023, 8028, 8029, 8030, 8033, 8037, 8051, 8052, 8055, 8056, 8065, 8066, 8074, 8431.

8016.   A02
How much power must a 24-volt generator furnish to a system which contains the following loads?

UNIT	RATING
One motor (75 percent efficient)	1/5 hp
Three position lights	20 watts each
One heating element	5 amp
One anticollision light	3 amp

(Note: 1 horsepower = 746 watts)

A — 402 watts.
B — 385 watts.
C — 450 watts.

8016. Answer C. JSGT 3-14 (AC 65-9A)
To solve this problem you must first calculate the power (P = EI) used by each unit.

One Motor (746 watts × 1/5 HP) ÷ 75% = 198.93 watts
Pos. Lights 3 lights × 20 watts = ...60.00 watts
Heating Elt. 24 volts × 5 amps = ...120.00 watts
Anticollision 24 volts × 3 amps = ...72.00 watts
The total required power output is ....450.93 watts

Answer (C), 450 watts, is the closest.

8017.   A02
A 12-volt electric motor has 1,000 watts input and 1 horsepower output. Maintaining the same efficiency, how much input power will a 24-volt, 1-horsepower electric motor require?
(Note: 1 horsepower = 746 watts)

A — 1,000 watts.
B — 2,000 watts.
C — 500 watts.

8017. Answer A. JSGT 3-14 (AC 65-9A)
As long as the same efficiency is maintained, a 1 HP motor requires 1,000 watts regardless of the system voltage. The advantage of using a higher system voltage is that less current is required and a smaller feed line can be used.

8019.   **A02**
A 1-horsepower, 24-volt dc electric motor that is 80 percent efficient requires 932.5 watts. How much power will a 1-horsepower, 12-volt dc electric motor that is 75 percent efficient require?
Note: 1 horsepower = 746 watts)

A — 932.5 watts.
B — 1,305.5 watts.
C — 994.6 watts.

8020.   **A02**
The potential difference between two conductors which are insulated from each other is measured in

A — volts.
B — amperes.
C — coulombs.

8023.   **A02**
(Refer to figure 4) How much power is being furnished to the circuit?

A — 575 watts.
B — 2,875 watts.
C — 2,645 watts.

8028.   **A03**
(Refer to figure 9) How many instruments (voltmeters and ammeters) are installed correctly?

A — Three.
B — One.
C — Two.

8019. Answer C. JSGT 3-15 (AC 65-9A)
If a motor operates at 100 percent efficiency it consumes 746 watts of energy for each horsepower developed. However, because there are always friction and heat loss, a motor is never 100 percent efficient and more than 746 watts is needed to produce 1 horsepower. To determine the number of watts required to produce 1 horsepower, divide the number of watts in 1 horsepower by the efficiency of the motor. A 1-horsepower motor that is 75 percent efficient requires 994.6 watts (746 watts ÷ .75 = 994.6 watts).

8020. Answer A. JSGT 3-4 (AC 65-9A)
Potential difference is one way of expressing voltage. Other terms used include Potential, Electromotive Force (EMF), Voltage Drop, and IR Drop.

8023. Answer C. JSGT 3-13 (AC 65-9A)
Because the circuit shown is purely resistive, you can calculate power by using the formula P = I × E. However, first you must determine the system voltage. Ohm's law states that volts (E) equals amperes (I) times resistance (R) or E = I × R. The circuit voltage is 115 volts (23 amps × 5 ohms = 115 volts). Now, use the power formula to calculate the amount of power being furnished. The answer is 2,645 watts (23 amps × 115 volts = 2,645 watts).

8028. Answer C. JSGT 3-23 (AC 65-9A)
Basic rules to follow are that ammeters are connected in series, voltmeters are connected in parallel, and proper polarity must be observed. In this circuit, the voltmeter in parallel with the light and the ammeter in series with the light and the battery are connected properly. The polarity on the second voltmeter is incorrect and the second ammeter is in parallel with the battery.

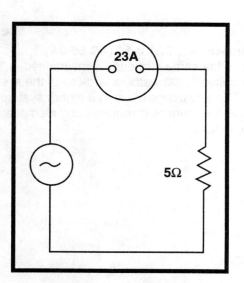

FIGURE 4.—Circuit Diagram.

Basic Electricity

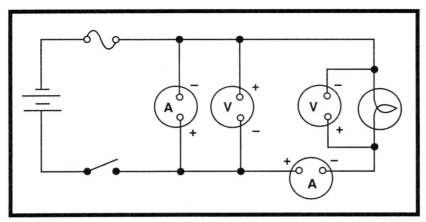

FIGURE 9.—Circuit Diagram.

8029. A03
The correct way to connect a test voltmeter in a circuit is

A — in series with a unit.
B — between source voltage and the load.
C — in parallel with a unit.

8030. A03
Which term means .001 ampere?

A — Microampere.
B — Kiloampere.
C — Milliampere.

8033. A03
.002KV equals

A — 20 volts.
B — 2.0 volts.
C — .2 volt.

8037. A04
What unit is used to express electrical power?

A — Volt.
B — Watt.
C — Ampere.

8029. Answer C. JSGT 3-23 (AC 65-9A)
Voltmeters are always connected in parallel with the unit being checked and proper polarity must be observed to prevent damage to the meter. It is handy to remember, measure voltage-across (parallel), measure current-through (series).

8030. Answer C. JSGT 3-5 (AC 65-9A)
Each of the terms listed here are common metric prefixes. You should be familiar with the values of all 3 prefixes listed in this question. Microampere = 0.000,001 ampere, Kiloampere = 1,000 amps, and Milliampere = .001 ampere.

8033. Answer B. JSGT 3-5 (AC 65-9A)
The metric prefix Kilo means 1,000. Therefore, .002KV is equal to 2 volts (.002 KV × 1,000 = 2.0 volts).

8037. Answer B. JSGT 3-5 (AC 65-9A)
The watt is used to express electrical power. The volt (answer A) is a measure of electromotive force and the ampere (answer C) is a measure of current flow.

8051. A04
Which of these will cause the resistance of a conductor to decrease?

A — Decrease the length or the cross-sectional area.
B — Decrease the length or increase the cross-sectional area.
C — Increase the length or decrease the cross-sectional area.

8052. A04
Through which material will magnetic lines of force pass the most readily?

A — Copper.
B — Iron.
C — Aluminum.

8055. A04
The voltage drop in a conductor of known resistance is dependent on

A — the voltage of the circuit.
B — only the resistance of the conductor and does not change with a change in either voltage or amperage.
C — the amperage of the circuit.

8056. A05
A thermal switch, as used in an electric motor, is designed to

A — close the integral fan circuit to allow cooling of the motor.
B — open the circuit in order to to allow cooling of the motor.
C — reroute the circuit to ground.

8065. A05
(Refer to figure 17) Which of the components is a potentiometer?

A — 5.
B — 3.
C — 11.

8066. A05
(Refer to figure 17) The electrical symbol represented at number 5 is a variable

A — inductor.
B — resistor.
C — capacitor.

8051. Answer B. JSGT 3-16 (AC 65-9A)
There are four factors which affect the resistance of a conductor. They are (1) the type of material used; (2) the length of the conductor; (3) the size of the cross sectional area; and (4) temperature. If you wish to decrease the resistance of a conductor, you can (1) select a different material, one that has a lower resistivity; (2) decrease the length of the conductor, the shorter the conductor the less the resistance; (3) increase the cross-sectional area of the conductor, a larger cross section results in a lower resistance; or (4) lower the temperature.

8052. Answer B. JSGT 3-10 (AC 65-9A)
The measure of the ease with which lines of magnetic flux pass through a material is measured in terms of permeability. The permeability scale is based on a perfect vacuum with air used as a reference and is given the permeability of one. Flux can travel through iron much easier than air or other materials because it has a permeability of approximately 7,000.

8055. Answer C. JSGT 3-14 (AC 65-9A)
Ohm's law states that two variables affect voltage. They are current and resistance. This can be seen in the formula $E = I \times R$. Since the resistance in an ordinary conductor is constant, the voltage drop in a conductor is dependant on the current flowing through the conductor.

8056. Answer B. JSGT 3-19 (AC 65-9A)
Thermal switches (circuit breakers) are incorporated in many motors to prevent damage in case of overheat. A thermal switch automatically opens the circuit when excess current heats an element in the switch.

8065. Answer B. JSGT 3-22 (AC 65-9A)
The component illustrated at 3 is a potentiometer. The component at 5 (answer A) is a variable capacitor, and the component at 11 (answer C) is an inductor or coil.

8066. Answer C. JSGT 3-22 (AC 65-9A)
The component illustrated at 5 is a variable capacitor, this symbol is created by drawing an arrow through the symbol for a capacitor to indicate that its capacitance is variable. An inductor (answer A) is illustrated at 11 and a resistor (answer B) is illustrated at 6 and 7.

Basic Electricity

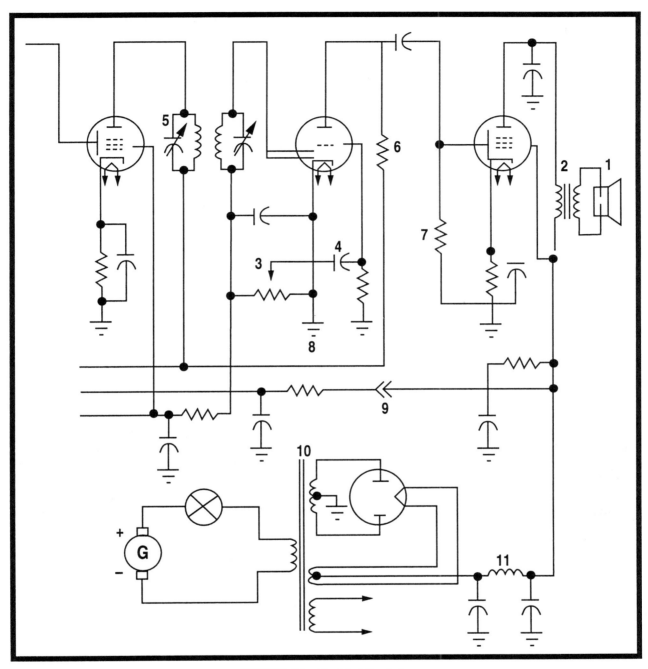

FIGURE 17.—Electrical Symbols.

8074.   **A05**
(Refer to figure 21) Which symbol represents a variable resistor?

A — 2.
B — 1.
C — 3.

8074. Answer A. JSGT 3-22 (AC 65-9A)
Selection 2 (answer A) illustrates a resistor with an arrow to indicate it is variable. Selection 1 (answer B) could be either a rheostat or potentiometer and selection 3 (answer C) is a tapped resistor.

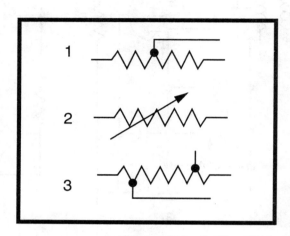

FIGURE 21.—Electrical Symbols.

8431.   **H03**
How much current does a 30-volt motor, 1/2 horsepower, 85 percent efficient draw from the bus? (Note: 1 horsepower = 746 watts)

A — 14.6 amperes.
B — 12.4 amperes.
C — 14.1 amperes.

8431. Answer A. JSGT 3-14 (AC 65-9A)
To solve this problem, you must use the formula I = P ÷ E, where (I) equals amps or current, (P) equals power or watts, and (E) equals volts. From the information given you can calculate that a 100 percent efficient 1/2 HP motor uses 373 watts of power (746 ÷ 2 = 373). However, the motor in this question is only 85 percent efficient. To determine the actual number of watts used to produce 1/2 HP, you must divide 373 watts by 85 percent. The actual power used is 438.8 watts (373 ÷ .85 = 438.8). Therefore, the current required is 14.6 amps (438.8 watts ÷ 30 volts = 14.6 amps)

Basic Electricity

SECTION B
DIRECT CURRENT

Section B of Chapter 3 discusses the fundamentals of direct current (DC) including basic terminology and the structure of direct current. The section continues by discussing series DC circuits, parallel DC circuits, and how DC is converted to AC. The following FAA Test questions pertain to Section B:

8015, 8018, 8021, 8025, 8026, 8027, 8031, 8032, 8034, 8035, 8036, 8038, 8039, 8042, 8043, 8044, 8045, 8046, 8047, 8049, 8050, 8053, 8054.

8015. A02
Which requires the most electrical power during operation? (Note: 1 horsepower = 746 watts)

A — A 12-volt motor requiring 8 amperes.
B — Four 30-watt lamps in a 12-volt parallel circuit.
C — Two lights requiring 3 amperes each in a 24-volt parallel system.

8015. Answer C. JSGT 3-28 (AC 65-9A)
Power is defined in terms of watts (P), and is the product of volts (E) and amps (I). Each light in choice (C) requires 72 watts of power (3 amps × 24 volts = 72 watts). Since there are 2 lights, a total of 144 watts is required (72 watts × 2 lights = 144 watts). Answer (A) requires 96 watts of power (12 volts × 8 amps = 96 watts), and answer (B) requires 120 watts (4 lamps × 30 watts = 120 watts). Therefore, answer (C) is the correct answer since it requires the most electrical power.

8018. A02
How many amperes will a 28-volt generator be required to supply to a circuit containing five lamps in parallel, three of which have a resistance of 6 ohms each and two of which have a resistance of 5 ohms each?

A — 1.11 amperes.
B — 1 ampere.
C — 25.23 amperes.

8018. Answer C. JSGT 3-27 (AC 65-9A)
This problem requires application of the formula for determining total resistance in a parallel circuit:

$$\frac{1}{\frac{1}{R_1}+\frac{1}{R_2}+\frac{1}{R_3}}$$

The total resistance is 1.11 ohms. Once you know the voltage and resistance in a circuit, you can determine the amperes required by applying Ohm's law (I = E ÷ R). The answer is 25.23 amperes (28 volts ÷ 1.11 ohms = 25.23 amps).

8021. A03
A 24-volt source is required to furnish 48 watts to a parallel circuit consisting of four resistors of equal value. What is the voltage drop across each resistor?

A — 12 volts.
B — 3 volts.
C — 24 volts.

8021. Answer C. JSGT 3-27 (AC 65-9A)
Kirchhoff's law states that in a parallel circuit, voltage remains constant and amperes vary across each resistance. Therefore, the voltage drop measured at each resistance in a 24-volt circuit must equal the applied voltage of 24 volts.

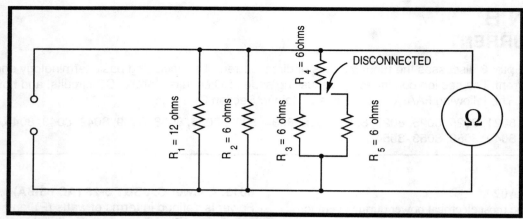

FIGURE 6.—Circuit Diagram.

8025. A03
(Refer to figure 6) If resistor R_5 is disconnected at the junction of R_4 and R_3 as shown, what will the ohmmeter read?

A — 2.76 ohms.
B — 3 ohms.
C — 12 ohms.

8025. Answer B. JSGT 3-29 (AC 65-9A)
With resistor R_5 disconnected, the ohmmeter reads the resistance of the remaining four resistors. R_3 and R_4 are in series and, therefore, can be combined, resulting in a total of 12 ohms. This 12 ohm total is in parallel with the remaining two resistors. The formula for parallel resistances is now used to calculate the total resistance of 3 ohms.

$$\frac{1}{\frac{1}{12}+\frac{1}{6}+\frac{1}{12}} = 3 \text{ ohms}$$

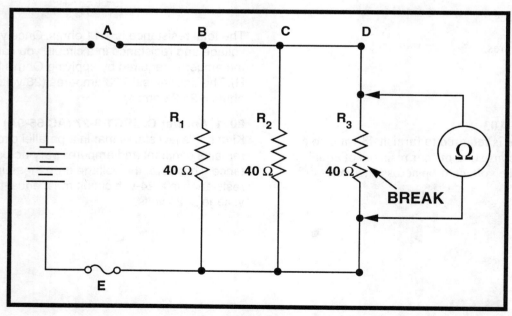

FIGURE 7.—Circuit Diagram.

8026. A03
(Refer to figure 7) If resistor R_3 is disconnected at terminal D, what will the ohmmeter read?

A — Infinite resistance.
B — 10 ohms.
C — 20 ohms.

8026. Answer A. JSGT 3-27 (AC 65-9A)
By disconnecting resistor R_3 at terminal D, the flow is broken to the rest of the circuit, and the break in the resistor itself is identified by an infinite resistance reading on the ohmmeter (answer A). If resistor R_3 was not disconnected at terminal D, the ohmmeter would indicate the resistance of R_1 and R_2 which is 20 ohms (answer C).

Basic Electricity

8027.   A03
(Refer to figure 8) With an ohmmeter connected into the circuit as shown, what will the ohmmeter read?

A — 20 ohms.
B — Infinite resistance.
C — 10 ohms.

8027. Answer C. JSGT 3-27 (AC 65-9A)
Because of the break in resistor R_3, resistors R_1 and R_2 are the only resistances measured by the ohmmeter. These two resistors are connected in parallel and, therefore, their combined resistance can be calculated with the formula:

$$\frac{1}{\frac{1}{R_1} + \frac{1}{R_2}}$$

The total resistance of R_1 and R_2 is 10 ohms

$$\frac{1}{\frac{1}{20} + \frac{1}{20}} = 10 \text{ ohms}$$

and represents the value displayed on the meter.

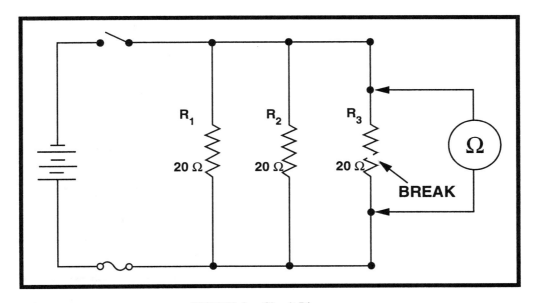

FIGURE 8.—Circuit Diagram.

8031.   A03
A cabin entry light of 10 watts and a dome light of 20 watts are connected in parallel to a 30-volt source. If the voltage across the 10-watt light is measured, it will be

A — equal to the voltage across the 20-watt light.
B — half the voltage across the 20-watt light.
C — one-third of the input voltage.

8031. Answer A. JSGT 3-27 (AC 65-9A)
In a parallel circuit, the voltage remains constant across each path while amperes vary. Therefore, the voltage across the 10-watt light is equal to the voltage across the 20-watt light.

8032.   A03
A 14-ohm resistor is to be installed in a series circuit carrying .05 ampere. How much power will the resistor be required to dissipate?

A — At least .70 milliwatt.
B — At least 35 milliwatts.
C — Less than .035 watt.

8032. Answer B. JSGT 3-26 (AC 65-9A)
Given resistance (R) and current (I), you can calculate the circuit voltage (E) using the formula E=IR. The circuit voltage equals .7 volts (.05 amps × 14 ohms = .7 volts). Once voltage is known, you can calculate power (P) with the formula P = IE. The power the resistor must dissipate is .035 watts or 35 milliwatts (.05 amps × .7 volts = .035 watts).

8034. A03
(Refer to figure 10) What is the measured voltage of the series-parallel circuit between terminals A and B?

A — 1.5 volts.
B — 3.0 volts.
C — 4.5 volts.

8034. Answer B. JSGT 3-29 (AC 65-9A)
Probably the easiest way to see how these batteries are connected is to redraw the circuit (be careful to observe proper polarity). The measured voltage is 3 volts.

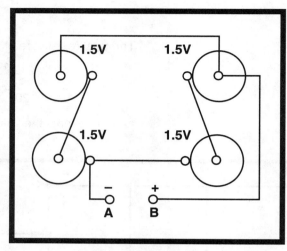

FIGURE 10.—Battery Circuit.

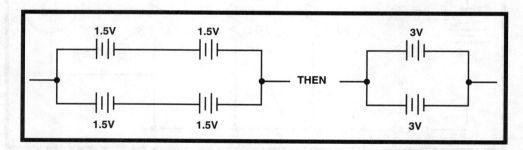

8035. A04
A 24-volt source is required to furnish 48 watts to a parallel circuit consisting of two resistors of equal value. What is the value of each resistor?
(Note: $R_T = E^2/P$)

A — 24 ohms.
B — 12 ohms.
C — 6 ohms.

8035. Answer A. JSGT 3-27 (AC 65-9A)
Using the formula given, you can calculate the total resistance of 12 ohms (24 volts² ÷ 48 watts = 12 ohms). The formula for determining the individual resistance of two like resistors connected in parallel is $r = R_T \times 2$, where (r) represents the resistance of each resistor and (R_T) represents the total resistance. The value of each resistor is 24 ohms (12 ohms × 2 = 24 ohms).

8036. A04
Which requires the most electrical power?
(Note: 1 horsepower = 746 watts)

A — Four 30-watt lamps arranged in a 12-volt parallel circuit.
B — A 1/5-horsepower, 24-volt motor which is 75 percent efficient.
C — A 24-volt anticollision light circuit consisting of two light assemblies which require 3 amperes each during operation.

8036. Answer C. JSGT 3-26 (AC 65-9A)
To answer this question, you must calculate the power requirements (wattage) of each unit listed.

Lamps.............................4 3 30 watts = 120W
24V motor ....(746 watts 3 1/10 HP) ÷ 75% = 99.47W
Anticollision ......24 volts 3 3 amps 3 2 assemblies = 144W

Answer (C) requires the most electrical power.

8038. A04

What is the operating resistance of a 30-watt light bulb designed for a 28-volt system?

A — 1.07 ohms.
B — 26 ohms.
C — 0.93 ohm.

8039. A04

Which statement is correct when made in reference to a parallel circuit?

A — The current is equal in all portions of the circuit.
B — The total current is equal to the sum of the currents through the individual branches of the circuit.
C — The current in amperes can be found by dividing the EMF in volts by the sum of the resistors in ohms.

8042. A04

If three resistors of 3 ohms, 5 ohms, and 22 ohms are connected in series in a 28-volt circuit, how much current will flow through the 3-ohm resistor?

A — 9.3 amperes.
B — 1.05 amperes.
C — 0.93 ampere.

8043. A04

A circuit has an applied voltage of 30 volts and a load consisting of a 10-ohm resistor in series with a 20-ohm resistor. What is the voltage drop across the 10-ohm resistor?

A — 10 volts.
B — 20 volts.
C — 30 volts.

8038. Answer B. JSGT 3-25 (AC 65-9A)

To determine resistance, Ohm's law states that resistance (R) equals volts (E) divided by amperes (I) or R = E ÷ I. Therefore, to solve this problem, amperes must be determined. Using the given information, you can calculate amps with the formula I = P ÷ E. The number of amperes in this circuit is 1.07 amps (30 watts ÷ 28 volts = 1.07 amps). With the number of amps known you can calculate the resistance. The operating resistance is 26.17 ohms (28 volts ÷ 1.07 = 26.17 ohms). Answer (B) is the closest.

8039. Answer B. JSGT 3-27 (AC 65-9A)

In a parallel circuit, the voltage remains constant through each unit and the current flow varies with each unit's resistance. However, Kirchhoff's law states that the current flowing to a point must equal the current flowing away from the point. Therefore, the total current flow in a parallel circuit is equal to the sum of the currents through each branch of the circuit.

8042. Answer C. JSGT 3-25 (AC 65-9A)

Kirchhoff's law states that in a series circuit, current remains constant and voltage varies across each resistor. To determine the current, use the formula I = E ÷ R. The total resistance in a series circuit is calculated by adding all the resistances. The total resistance in this circuit is 30 ohms (3 + 5 + 22 = 30 ohms). The total current within the circuit is .93 amps (28 volts ÷ 30 ohms = .93 amps). Since current remains constant, the 3 ohm resistor has .93 amps flowing through it.

8043. Answer A. JSGT 3-25 (AC 65-9A)

In a series circuit, the current is the same in all parts of the circuit, but the voltage drop varies with the resistance of each unit. However, before the voltage drop can be determined in this problem, you must calculate the total resistance and current. Total resistance equals 30 ohms (10 + 20 = 30). Now, use the formula I = E ÷ R to calculate the current flowing through the circuit. There is 1 amp flowing through the circuit (30 volts ÷ 30 ohms = 1 amp). To determine the voltage drop across the 10 ohm resistor use the formula $E_1 = I \times R_1$. The voltage drop across the 10 ohm resistor is 10 volts (1 amp × 10 ohms = 10 volts).

8044.   **A04**
(Refer to figure 11) Find the total current flowing in the wire between points C and D.

A — 6.0 amperes.
B — 2.4 amperes.
C — 3.0 amperes.

8044. Answer C. JSGT 3-27 (AC 65-9A)
Current flow in a parallel circuit varies with the resistance value of each branch. To determine the amount of current flowing from point C to point D you must first calculate the circuit's total resistance. This is done with the formula

$$\frac{1}{\frac{1}{8} + \frac{1}{10} + \frac{1}{40}}$$

The total resistance is 4 ohms. Now determine the total current flowing through the circuit using the formula I = E ÷ R. The total current is 6 amps (24 volts ÷ 4 ohms = 6 amps). Since this is a parallel circuit, once the current reaches point C a portion of it proceeds to R_1 and the rest to point D. To calculate the amount of current that flows between point C and D, you must calculate the amount of current flowing to R_1. To do this, use the formula I_{R1} = E ÷ R_1. The current drop across R_1 is 3 amps (24 volts ÷ 8 ohms = 3 amps). This leaves 3 amps to flow between points C and D (6 amps − 3 amps = 3 amps).

8045.   **A04**
(Refer to figure 11) Find the voltage across the 8-ohm resistor.

A — 8 volts.
B — 20.4 volts.
C — 24 volts.

8045. Answer C. JSGT 3-27 (AC 65-9A)
In a parallel circuit, the voltage remains constant across each resistor. This means the voltage across the 8-ohm resistor is equal to the source voltage of 24 volts.

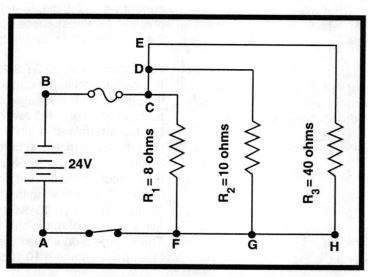

FIGURE 11.—Circuit Diagram.

8046.   A04
(Refer to figure 12) Find the total resistance of the circuit.

A — 16 ohms.
B — 2.6 ohms.
C — 21.2 ohms.

8046. Answer C. JSGT 3-29 (AC 65-9A)
When solving complex circuit problems for total resistance, begin by solving the parallel branches first. The formulas given can be used if you begin by solving for R_a and then continue by solving for R_b, R_c, and R_t respectively. Solving for R_a combines resistors R_4 and R_5. The combined resistance of R_4 and R_5 is 4 ohms. Since R_2 and $R_{4,5}$ are in series, add these resistances. This is demonstrated in the formula $R_b = R_a + R_2$. The combined resistance of R_2, and $R_{4,5}$ is 16 ohms (4 ohms + 12 ohms = 16 ohms). Now, using the formula to solve for R_c, combine the resistances of R_3 and $R_{2,4,5}$ which are in parallel. The combined resistance is 3.2 ohms. The final step is to add the remaining two resistances (R_1 and $R_{2,3,4,5}$) which are in series. The total resistance in the circuit is 21.2 ohms (18 ohms + 3.2 ohms = 21.2 ohms).

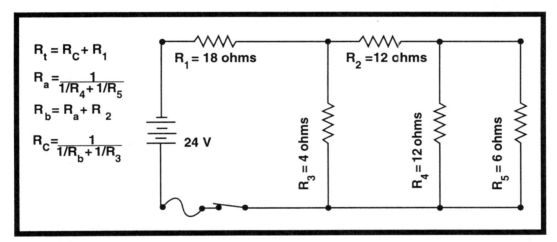

FIGURE 12.—Circuit Diagram.

8047.   A04
Which is correct in reference to electrical resistance?

A — Two electrical devices will have the same combined resistance if they are connected in series as they will have if connected in parallel.
B — If one of three bulbs in a parallel lighting circuit is removed, the total resistance of the circuit will become greater.
C — An electrical device that has a high resistance will use more power than one with a low resistance with the same applied voltage.

8047. Answer B. JSGT 3-27 (AC 65-9A)
In a parallel circuit, the greater the number of resistors, the less the total resistance. If you remove a resistor from a parallel circuit, the total resistance in the circuit goes up. This is the same as plugging too many appliances into a single outlet in your house causing the circuit breaker to pop. The total resistance goes down, the current flow goes up, and the breaker overloads.

8049.   **A04**

(Refer to figure 13) Determine the total current flow in the circuit.

A — 0.2 ampere.
B — 1.4 amperes.
C — 0.8 ampere.

8049. Answer B. JSGT 3-27 (AC 65-9A)
The total current flow in a parallel circuit is equal to the sum of the current flowing through each branch of the circuit. You can calculate the current in each branch using the formula I = E ÷ R. Resistor 1 has .4 amps of current (12V ÷ 30 ohms = .4 amps). Resistor 2 has .2 amps of current (12V ÷ 60 ohms = .2 amps). And, resistor 3 has .8 amps of current (12V ÷ 15 ohms = .8 amps). This results in a total current flow of 1.4 amps (.4 + .2 + .8 = 1.4).

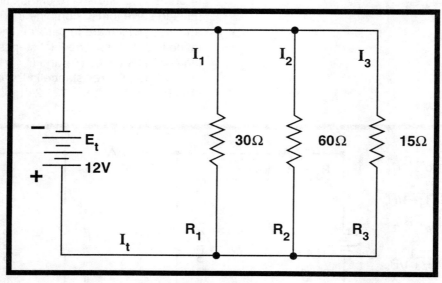

FIGURE 13.—Circuit Diagram.

8050.   **A04**

(Refer to figure 14) The total resistance of the circuit is

A — 25 ohms.
B — 35 ohms.
C — 17 ohms.

8050. Answer C. JSGT 3-29 (AC 65-9A)
When presented with a series-parallel circuit and asked to calculate the total resistance, begin at the point farthest from the power source and work back to the power. Here, you can begin by finding the total resistance of R_2, R_3, and R_4 which are in parallel. Their combined resistance is 2 ohms.

$$\frac{1}{\frac{1}{4} + \frac{1}{6} + \frac{1}{12}} = 2 \text{ ohms}$$

Now, all resistances are in series. To determine the total resistance of a series circuit add all the resistances. The circuit's total resistance is 17 ohms (5 ohms + 2 ohms + 10 ohms = 17 ohms).

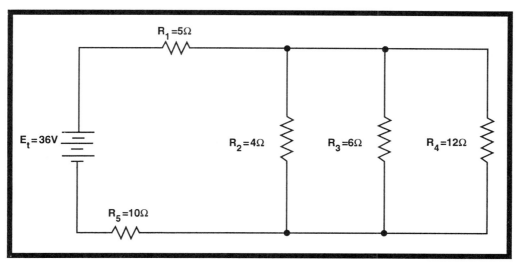

FIGURE 14.—Circuit Diagram.

8053. A04
A 48-volt source is required to furnish 192 watts to a parallel circuit consisting of three resistors of equal value. What is the value of each resistor?

A — 36 ohms.
B — 4 ohms.
C — 12 ohms.

8053. Answer A. JSGT 3-27 (AC 65-9A)
To solve this problem, you must use three Ohm's law formulas. First, determine the total current in the circuit using the formula I = P ÷ E. The current is 4 amps (192 watts ÷ 48 volts = 4 amps). Now, calculate the circuit's total resistance using the formula R = E ÷ I. Total resistance is 12 ohms (48 volts ÷ 4 amps = 12 ohms). To determine the resistance of each resistor use the formula r = R_T × n where r = the resistance of each resistor, R_T = total resistance, and n = the number of resistors in the circuit. The value of each resistor is 36 ohms (12 ohms × 3 = 36 ohms).

8054. A04
Which is correct concerning a parallel circuit?

A — Total resistance will be smaller than the smallest resistor.
B — Total resistance will decrease when one of the resistances is removed.
C — Total voltage drop is the same as the total resistance.

8054. Answer A. JSGT 3-28 (AC 65-9A)
According to Kirchhoff's law, the total resistance in a parallel circuit is always less than the smallest resistor. This can also be derived from the formula

$$R_T = \frac{1}{\frac{1}{R_1} + \frac{1}{R_2} + \frac{1}{R_3}}$$

SECTION C
BATTERIES

Section C looks at the primary and secondary cell batteries found in aviation. The main emphasis in this section falls on the characteristics and principles associated with lead-acid and nickel-cadmium batteries. This includes operational characteristics as well as how to service each type of battery. The following FAA Test questions are discussed:

8085, 8086, 8087, 8088, 8089, 8090, 8091, 8092, 8093, 8094, 8095, 8096, 8097, 8098, 8099, 8100, 8101, 8102, 8351.

8085.   A06
A lead-acid battery with 12 cells connected in series (no-load voltage = 2.1 volts per cell) furnishes 10 amperes to a load of 2-ohms resistance. The internal resistance of the battery in this instance is

A — 0.52 ohm.
B — 2.52 ohms.
C — 5.0 ohms.

8085. Answer A. JSGT 3-37 (AC 65-9A)
To calculate internal resistance, subtract the closed circuit voltage from the no-load voltage and divide by closed circuit current. The no-load voltage is 25.2 volts (12 cells × 2.1 volts = 25.2 volts), the closed circuit voltage is 20 volts (10 amps × 2 ohms = 20 volts), and the closed circuit current is 10 amps. The battery's internal resistance is .52 ohms (25.2 volts – 20 volts) ÷ 10 amps = .52 ohms.

8086.   A06
If electrolyte from a lead-acid battery is spilled in the battery compartment, which procedure should be followed?

A — Apply boric acid solution to the affected area followed by a water rinse.
B — Rinse the affected area thoroughly with clean water.
C — Apply sodium bicarbonate solution to the affected area followed by a water rinse.

8086. Answer C. JSGT 3-39 (AC 43.13-1A)
Sodium bicarbonate (baking soda) is used to neutralize the electrolyte from a lead-acid battery. If electrolyte is spilled, you should immediately apply sodium bicarbonate and rinse the contaminated area with water.

8087.   A06
Which statement regarding the hydrometer reading of a lead-acid storage battery electrolyte is true?

A — The hydrometer reading does not require a temperature correction if the electrolyte temperature is 80°F.
B — A specific gravity correction should be subtracted from the hydrometer reading if the electrolyte temperature is above 20°F.
C — The hydrometer reading will give a true indication of the capacity of the battery regardless of the electrolyte temperature.

8087. Answer A. JSGT 3-37 (AC 65-9A)
A hydrometer accurately measures the specific gravity of battery electrolyte when it is at or near 80°F. For temperatures above 90°F or below 70°F, it is necessary to apply a correction factor. Some hydrometers are equipped with a scale inside the tube. With other hydrometers it is necessary to refer to a chart provided by the manufacturer.

8088.   A06
A fully charged lead-acid battery will not freeze until extremely low temperatures are reached because

A — the acid is in the plates, thereby increasing the specific gravity of the solution.
B — most of the acid is in the solution.
C — increased internal resistance generates sufficient heat to prevent freezing.

8088. Answer B. JSGT 3-36 (AC 65-9A)
When a lead-acid battery is fully charged, the electrolyte contains a high concentration of sulfuric acid. This high concentration of acid raises the specific gravity of the electrolyte and lowers the freezing point. A fully charged lead-acid battery has a freezing point of –80°F to –90°F.

8089.   A06
What determines the amount of current which will flow through a battery while it is being charged by a constant voltage source?

A — The total plate area of the battery.
B — The state-of-charge of the battery.
C — The ampere-hour capacity of the battery.

8089. Answer B. JSGT 3-40 (AC 65-9A)
When a battery is discharged, its low voltage allows a large amount of current to flow into the battery. As the battery charges and the voltage rises, the current flow decreases.

8090.   A06
Which of the following statements is/are generally true regarding the charging of several aircraft batteries together?
1. Batteries of different voltages (but similar capacities) can be connected in series with each other across the charger, and charged using the constant current method.
2. Batteries of different ampere-hour capacity and same voltage can be connected in parallel with each other across the charger, and charged using the constant voltage method.
3. Batteries of the same voltage and same ampere-hour capacity must be connected in series with each other across the charger, and charged using the constant current method.

A — 3.
B — 2 and 3.
C — 1 and 2.

8090. Answer C. JSGT 3-40 (AC 65-9A)
Choices 1 and 2 are correct. One of the nice features of constant-current chargers is that they can be used to charge batteries of different voltage at the same time. To do this, the batteries must be connected in series so that the current supply is the same to each battery (current is the same in all parts of a series circuit). The constant-current charge requires more time to charge a battery fully, as well as additional monitoring to avoid overcharging. You can also charge multiple batteries using a constant-voltage charger. When doing this, the batteries may have different ampere-hour ratings but they must have the same voltage and be connected in parallel.

8091.   A06
The method used to rapidly charge a nickel-cadmium battery utilizes

A — constant current and constant voltage.
B — constant current and varying voltage.
C — constant voltage and varying current.

8091. Answer C. JSGT 3-43 (AC 65-9A)
Whenever a battery is charged rapidly, it is a constant-voltage type charge. When using a constant-voltage charge on a nickel-cadmium battery the voltage remains constant and the current decreases as the battery charges.

8092.   A06
If an aircraft ammeter shows a full charging rate, but the battery remains in a discharged state, the most likely cause is

A — an externally shorted battery .
B — an internally shorted battery.
C — a shorted generator field circuit.

8092. Answer B. JSGT 3-41 (AC 65-9A)
If a battery can no longer hold a charge, it is typically a good indicator that the battery is shorted internally.

8093.   A06
Which condition is an indication of improperly torqued cell link connections of a nickel-cadmium battery?

A — Light spewing at the cell caps.
B — Toxic and corrosive deposit of potassium carbonate crystals.
C — Heat or burn marks on the hardware.

8093. Answer C. JSGT 3-43 (AC 65-9A)
If the cell link connections are not properly torqued, arcing and overheating may occur. You can identify this condition by the presence of heat or blue marks on the hardware.

8094. A06
The presence of small amounts of potassium carbonate deposits on the top of nickel-cadmium battery cells that have been in service for a time is an indication of

A — normal operation.
B — excessive gassing.
C — excessive plate sulfation.

8095. A06
The servicing and charging of nickel-cadmium and lead-acid batteries together in the same service area is likely to result in

A — normal battery service life.
B — increased explosion and/or fire hazard.
C — contamination of both types of batteries.

8096. A06
The electrolyte of a nickel-cadmium battery is the lowest when the battery is

A — being charged.
B — in a discharged condition.
C — under a heavy load condition.

8097. A06
The end-of-charge voltage of a 19-cell nickel-cadmium battery, measured while still on charge,

A — must be 1.2 to 1.3 volts per cell.
B — must be 1.4 volts per cell.
C — depends upon its temperature and the method used for charging.

8098. A06
Nickel-cadmium batteries which are stored for a long period of time will show a low liquid level because

A — of the decrease in the specific gravity of the electrolyte.
B — electrolyte evaporates through the vents.
C — electrolyte becomes absorbed into the plates.

8099. A06
How can the state-of-charge of a nickel-cadmium battery be determined?

A — By measuring the specific gravity of the electrolyte.
B — By a measured discharge.
C — By the level of the electrolyte.

8094. Answer A. JSGT 3-42 (AC 65-9A)
Most nickel-cadmium batteries develop an accumulation of white, potassium carbonate powder on top of the cells during normal operation. If there is an excessive amount, check the voltage regulator and the level of electrolyte in the cells.

8095. Answer C. JSGT 3-42 (AC 65-9A)
A separate storage and maintenance area should be provided for nickel-cadmium and lead acid batteries. The electrolyte used in lead-acid batteries is chemically opposite of that used in nickel-cadmium batteries. Any electrolyte transfer from one type to the other will result in contamination.

8096. Answer B. JSGT 3-42 (AC 65-9A)
During the discharge of a nickel-cadmium battery, the plates absorb a quantity of electrolyte. During recharge, the plates release the electrolyte and the level of the electrolyte rises. When fully charged, the electrolyte level of a nickel-cadmium battery is at its highest.

8097. Answer C. JSGT 3-44 (AC 65-9A)
The end of charge voltage, measured while the cell is on charge, depends upon its temperature and the method used for charging it.

8098. Answer C. JSGT 3-43 (AC 65-9A)
When a nickel-cadmium battery is stored for a long period of time, it may lose some or all of its charge. When this happens, the electrolyte is absorbed into the plates and the electrolyte level in the cell drops.

8099. Answer B. JSGT 3-43 (AC 65-9A)
Since the electrolyte of a nickel-cadmium battery does not react chemically with the cell plates, the specific gravity of the electrolyte does not change appreciably. For this reason, you cannot use a hydrometer to determine the state of charge in a nickel-cadmium battery. The only way to determine the condition of a nickel-cadmium battery is to fully charge it, then discharge it at a specified rate and measure its amp-hour capacity.

8100.   A06
What may result if water is added to a nickel-cadmium battery when it is not fully charged?

A — Excessive electrolyte dilution.
B — Excessive spewing is likely to occur during the charging cycle.
C — No adverse effects since water may be added anytime.

8100. Answer B. JSGT 3-43 (AC 65-9A)
When discharged, the plates of a nickel-cadmium battery absorb a quantity of the electrolyte. On recharge, the plates release the electrolyte and the level of electrolyte rises. When fully charged, the electrolyte is at its highest level. Therefore, water should be added only when the battery is fully charged. If water is added to a nickel-cadmium battery that is not fully charged, excessive spewing will occur during the charging cycle.

8101.   A06
In nickel-cadmium batteries, a rise in cell temperature

A — causes an increase in internal resistance.
B — causes a decrease in internal resistance.
C — increases cell voltage.

8101. Answer B. JSGT 3-41 (AC 65-9A)
The nickel-cadmium battery has a very low internal resistance. However, if a nickel-cadmium battery is subjected to high temperatures, the cellophane-like material that separates the plates begins to breakdown. The breakdown of this material decreases the battery's internal resistance further. Therefore, as cell temperature in a nickel-cadmium battery increases, internal resistance decreases.

8102.   A06
When a charging current is applied to a nickel-cadmium battery, the cells emit gas only

A — toward the end of the charging cycle.
B — when the electrolyte level is low.
C — if they are defective.

8102. Answer A. JSGT 3-42 (AC 65-9A)
A nickel-cadmium cell emits gas during the end of the charging cycle. The gas is caused by decomposition of the water in the electrolyte into hydrogen at the negative plates and oxygen at the positive plates. Caution should be observed as this gas is explosive. Proper ventilation must be provided while charging batteries.

8351.   G01
Nickel-cadmium battery cases and drain surfaces which have been affected by electrolyte should be neutralized with a solution of

A — boric acid.
B — sodium bicarbonate.
C — potassium hydroxide.

8351. Answer A. JSGT 3-42 (AC 43.13-1A)
Nickel-cadmium battery electrolyte is a strong base and, therefore, it must be neutralized by using an acid. A boric acid solution is the standard for this purpose. Sodium bicarbonate (answer B) is used to neutralize the electrolyte from lead acid batteries.

SECTION D
ALTERNATING CURRENT

Section D builds on the electrical theory discussed in prior sections to introduce the principles of alternating current (AC). Once you are familiar with the basic terminology and theory associated with alternating current, the section continues by discussing purely resistive AC circuits, as well as the principles of inductance and capacitance. The section closes by looking at three-phase alternating current and how AC is converted to DC. FAA Test questions that apply to Section D include:

   8001, 8002, 8003, 8004, 8005, 8006, 8007, 8008, 8009, 8010, 8011, 8012, 8013, 8014, 8022, 8024, 8041, 8048.

8001.   A01
The working voltage of a capacitor in an ac circuit should be

A — equal to the highest applied voltage.
B — at least 20 percent greater than the highest applied voltage.
C — at least 50 percent greater than the highest applied voltage.

8001. Answer C. JSGT 3-57 (AC 65-9A)
The working voltage of the capacitor is the maximum voltage that can be steadily applied without danger of arc-over, and depends on the type and thickness of the dielectric. When installing a capacitor in a circuit, the working voltage should be at least 50 percent greater than the highest applied voltage.

8002. A01

The term that describes the combined restive forces in an ac circuit is

A — resistance.
B — reactance.
C — impedance.

8003. A01

The basis for transformer operation in the use of alternating current is mutual

A — inductance.
B — capacitance.
C — reactance.

8004. A01

The opposition offered by a coil to the flow of alternating current is called (disregard resistance)

A — impedance.
B — reluctance.
C — inductive reactance.

8005. A01

An increase in which of the following factors will cause an increase in the inductive reactance of a circuit?

A — Inductance and frequency.
B — Resistance and voltage.
C — Resistance and capacitive reactance.

8006. A01

(Refer to figure 1.) When different rated capacitors are connected in series in a circuit, the total capacitance is

A — less than the capacitance of the lowest rated capacitor.
B — greater than the capacitance of the highest rated capacitor.
C — equal to the sum of all the capacitances.

8007. A01

In an ac circuit, the effective voltage is

A — equal to the maximum instantaneous voltage.
B — greater than the maximum instantaneous voltage.
C — less than the maximum instantaneous voltage.

8002. Answer C. JSGT 3-62 (AC 65-9A)

The flow of current in an AC circuit is opposed by three things: resistance, inductive reactance, and capacitive reactance. The combined effect of these three elements is known as impedance, and is represented by the letter Z. Impedance is obtained by finding the vector sum of the three oppositions.

8003. Answer A. JSGT 3-54 (AC 65-9A)

A basic transformer consists of two coils of wire. When alternating current flows through one coil, the changing lines of flux radiate out and cut across the second coil. Anytime lines of flux cut across another conductor, they induce a voltage in that conductor even though there is no electrical connection between the two. This is known as mutual inductance and is the basis for transformer operation.

8004. Answer C. JSGT 3-51 (AC 65-9A)

When alternating current flows through a coil of wire, a voltage is induced in the wire in the opposite direction of the applied voltage. This counter-EMF opposes the flow of current through the coil and is called inductive reactance.

8005. Answer A. JSGT 3-53 (AC 65-9A)

Inductive reactance is calculated with the formula:

$$X_L = 2\pi f L$$

If all other circuit values remain constant, the greater the inductance (L), the greater the inductive reactance. Furthermore, as the frequency (f) increases, inductive reactance also increases. Therefore, inductive reactance is directly proportional to the circuit inductance and frequency and an increase in either results in an increase in inductive reactance.

8006. Answer A. JSGT 3-58 (AC 65-9A)

When capacitors are connected in series, the total capacitance of the circuit is less than that of any single capacitor. This can be seen in the formula used to calculate total capacitance in a series circuit;

$$C_T = \frac{1}{\frac{1}{C_1} + \frac{1}{C_2} + \frac{1}{C_3}}$$

8007. Answer C. JSGT3-4-46 (AC 65-9A)

The effective voltage of alternating current is the same as the voltage of a direct current which produces the same heating effect. The effective voltage is always less than the maximum instantaneous voltage of the AC. Effective voltage in an AC circuit is also known as root mean squared, or RMS voltage and is calculated by multiplying .707 times the maximum instantaneous voltage.

Basic Electricity

3-21

$$C_T = \frac{1}{1/C_1 + 1/C_2 + 1/C_3 \ldots}$$

FIGURE 1.—Equation.

8008. A01
The amount of electricity a capacitor can store is directly proportional to the

A — distance between the plates and inversely proportional to the plate area.
B — plate area and is not affected by the distance between the plates.
C — plate area and inversely proportional to the distance between the plates.

8009. A01
(Refer to figure 2) What is the total capacitance of a certain circuit containing three capacitors with capacitances of .02 microfarad, .05 microfarad, and .10 microfarad, respectively?

A — 5.88 µF.
B — 0.125 pF.
C — .0125 µF.

8008. Answer C. JSGT 3-57 (AC 65-9A)
The capacity of a capacitor is affected by three variables: the area of the plates, the distance between the plates, and the dielectric constant of the material between the plates. The capacity is directly proportional to the plate area and inversely proportional to the distance between the plates. In other words, if the plate area increases the capacity increases, and if the distance between the plates increases total capacity decreases.

8009. Answer C. JSGT 3-58 (AC 65-9A)
To calculate total capacitance in a series circuit, use the formula in figure 2. The total capacitance is .0125 µF.

$$\frac{1}{\frac{1}{.02} + \frac{1}{.05} + \frac{1}{.01}} = .0125$$

Remember, total capacitance in a series circuit is less than any single capacitor.

$$C_T = \frac{1}{1/C_1 + 1/C_2 + 1/C_3}$$

FIGURE 2.—Equation.

8010. A01
Unless otherwise specified, any values given for current or voltage in an ac circuit are assumed to be

A — instantaneous values.
B — effective values.
C — maximum values.

8010. Answer B. JSGT 3-47 (AC 65-9A)
In the study of alternating current, all values given for current and voltage are assumed to be effective values unless otherwise specified and, in practice, only the effective values of voltage and current are used. AC voltmeters and ammeters measure the effective value.

8011.　　　　**A01**
When different rated capacitors are connected in parallel in a circuit, the total capacitance is
(Note: $C_T = C_1 + C_2 + C_3 ...$)

A — less than the capacitance of the lowest rated capacitor.
B — equal to the capacitance of the highest rated capacitor.
C — equal to the sum of all the capacitances.

8011. Answer C. JSGT 3-58 (AC 65-9A)
As seen in the formula given, when capacitors are connected in parallel, the total capacitance is equal to the sum of all the capacitances. Connecting capacitors in parallel gives the same effect as adding the areas of their plates.

8012.　　　　**A01**
When inductors are connected in series in a circuit, the total inductance is (where the magnetic fields of each inductor do not affect the others)
(Note: $L_T = L_1 + L_2 + L_3 ...$)

A — less than the inductance of the lowest rated inductor.
B — equal to the inductance of the highest rated inductor.
C — equal to the sum of the individual inductances.

8012. Answer C. JSGT 3-52 (AC 65-9A)
As seen in the formula given, when inductors are connected in series, the total inductance is equal to the sum of all the inductances.

8013.　　　　**A01**
(Refer to figure 3) When more than two inductors of different inductances are connected in parallel in a circuit, the total inductance is

A — less than the inductance of the lowest rated inductor.
B — equal to the inductance of the highest rated inductor.
C — equal to the sum of the individual inductances.

8013. Answer A. JSGT 3-52 (AC 65-9A)
Total inductance in a parallel circuit is calculated using the formula in figure 3. As you can see, total inductance in a parallel circuit equals the reciprocal sum of the reciprocal of the inductances. This means that total inductance in a parallel circuit is always less than the inductance of the lowest rated inductor.

$$L_T = \frac{1}{1/L_1 + 1/L_2 + 1/L_3 \ldots}$$

FIGURE 3.—Equation.

8014.　　　　**A01**
What is the total capacitance of a certain circuit containing three capacitors with capacitances of .25 microfarad, .03 microfarad, and .12 microfarad, respectively?
(Note: $C_T = C_1 + C_2 + C_3$)

A — .4 µF.
B — .04 pF.
C — .04 µF.

8014. Answer A. JSGT 3-58 (AC 65-9A)
As seen in the formula given, capacitances in parallel are additive. The total capacitance equals .4 µF (.25 + .03 + 0.12 = .4).

Basic Electricity

8022.   A02

When calculating power in a reactive or inductive ac circuit, the true power is

A — more than the apparent power.
B — less than the apparent power in a reactive circuit and more than the apparent power in an inductive circuit.
C — less than the apparent power.

8024.   A02

(Refer to figure 5) What is the impedance of an ac-series circuit consisting of an inductor with a reactance of 10 ohms, a capacitor with a reactance of 4 ohms, and a resistor with a resistance of 8 ohms?

A — 22 ohms.
B — 5.29 ohms.
C — 10 ohms.

8022. Answer C. JSGT 3-48 (AC 65-9A)
True power equals voltage times the portion of current that is in phase with the voltage. Apparent power, on the other hand, equals voltage times total current in and out of phase with the voltage. When capacitance or inductance is added to a circuit, the current and voltage are not exactly in phase. Therefore, true power in a reactive or inductive ac circuit is always less than the apparent power.

8024. Answer C. JSGT 3-63 (AC 65-9A)
This is an application of the formula given for computing impedance (figure 5). Since reactance values are already in ohms, plug the values given into the formula and perform the required calculations. The answer is 10 ohms.

$$\sqrt{8^2 + (10-4)^2} = 10 \text{ ohms}$$

$$Z = \sqrt{R^2 + (X_L - X_C)^2}$$

Z = Impedance
R = Resistance
X_L = Inductance Reactance
Z_C = Capacitive Reactance

FIGURE 5.—Formula.

8041.   A04

Transfer of electrical energy from one conductor to another without the aid of electrical connections

A — is called induction.
B — is called airgap transfer.
C — will cause excessive arcing and heat, and as a result is impractical.

8048.   A04

What happens to the current in a voltage step-up transformer with a ratio of 1 to 4?

A — The current is stepped down by a 1 to 4 ratio.
B — The current is stepped up by a 1 to 4 ratio.
C — The current does not change.

8041. Answer A. JSGT 3-54 (AC 65-9A)
Anytime a wire passes through a magnetic field, electrical energy is induced into the wire. Therefore, if you were to arrange two conductors (not connected electrically) such that the magnetic field surrounding one cuts through the other conductor, a current would be induced in the second conductor. This process is called induction and is the principle by which transformers operate.

8048. Answer A. JSGT 3-54 (AC 65-9A)
A transformer cannot generate power. Therefore, if a transformer steps up the voltage, it must step down the current by the same ratio. This is evident in the formula for power ($P = I \times E$). If voltage increases current must decrease, and if voltage decreases current must increase.

SECTION E
ELECTRON CONTROL DEVICES

Section E of Chapter 3 introduces semiconductor theory. It begins by looking at the principles of vacuum tubes and then applies those principles to modern semiconductor devices such as diodes, transistors, and logic gates. FAA Test questions pertaining to this section include:

    8040, 8075, 8076, 8077, 8078, 8079, 8080, 8081, 8082, 8083, 8084.

8040. A04
Diodes are used in electrical power circuits primarily as

A — cutout switches.
B — rectifiers.
C — relays.

8075. A05
In a P-N-P transistor application, the solid state device is turned on when the

A — base is negative with respect to the emitter.
B — base is positive with respect to the emitter.
C — emitter is negative with respect to the base.

8076. A05
In an N-P-N transistor application, the solid state device is turned on when the

A — emitter is positive with respect to the base.
B — base is negative with respect to the emitter.
C — base is positive with respect to the emitter.

8077. A05
Typical application for zener diodes is as

A — full-wave rectifiers.
B — half-wave rectifiers.
C — voltage regulators.

8040. Answer B. JSGT 3-70 (AC 65-9A)
The most important characteristic of a diode is that it permits current to flow in one direction only. The effect is like an electron check valve that permits flow in one direction but blocks any attempt to flow in the opposite direction. This characteristic permits diodes to be used to rectify AC to DC.

8075. Answer A. JSGT 3-77 (AC 65-9A)
To turn on a transistor, a small amount of current must flow into the base and the emitter-base must be forward-biased. A P-N-P transistor is forward-biased when the base is negative with respect to the emitter.

8076. Answer C. JSGT 3-77 (AC 65-9A)
To turn on a transistor, a small amount of current must flow into the base and the emitter-base must be forward-biased. An N-P-N transistor is forward-biased when the base is positive with respect to the emitter.

8077. Answer C. JSGT 3-76 (AC 65-9A)
In a zener diode, current will not flow through the diode until a specific voltage (zener voltage) is reached. When a zener diode is used in a circuit, current bypasses the diode until the zener voltage is obtained. However, once the zener voltage is exceeded, the additional voltage is dissipated through the diode. Because of this feature, zener diodes are commonly used in voltage regulators.

8078.　　　A05
(Refer to figure 22) Which illustration is correct concerning bias application and current flow?

A — 1.
B — 2.
C — 3.

8078. Answer A. JSGT 3-77 (AC 65-9A)
The first two symbols represent N-P-N transistors, while the third symbol represents a P-N-P transistor. In order for current to flow in a transistor, the emitter-base must be forward-biased. Forward-biasing in an N-P-N transistor requires the base to be positive with respect to the emitter. On the other hand, forward-biasing in a P-N-P transistor requires the base to be negative with respect to the emitter. When correct biasing exists in a transistor, current flows in the direction of the current flow arrow. In selection 1 (answer A) the base of the N-P-N transistor is positive with respect to the emitter indicating forward-biasing. In this configuration, emitter-base current flows so collector-emitter current flows in the direction of the current flow arrow. In selection 2 and 3 (answers B and C), the base and emitter have the same polarity so no emitter-base current exists. Therefore, there is no flow between the emitter and collector.

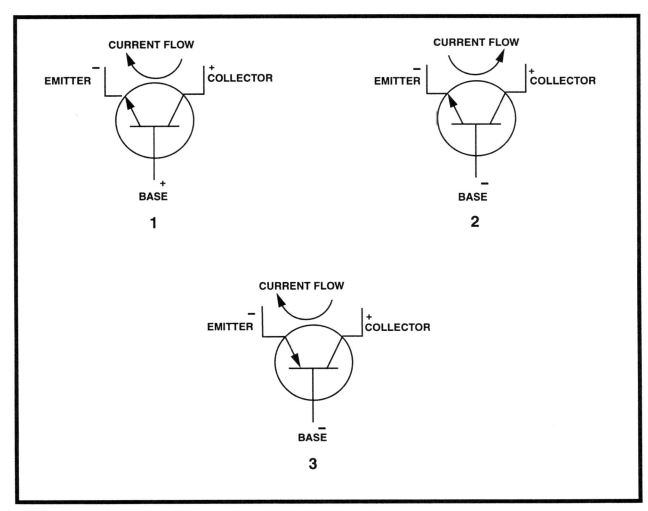

FIGURE 22.—Transistors.

8079.   A05
Forward biasing of a solid state device will cause the device to

A — conduct via zener breakdown.
B — conduct.
C — turn off.

8080.   A05
(Refer to figure 23) If an open occurs at R_1, the light

A — cannot be turned on.
B — will not be affected.
C — cannot be turned off.

8081.   A05
(Refer to figure 23) If R_2 sticks in the up position, the light will

A — be on full bright.
B — be very dim.
C — not illuminate.

8079. Answer B. JSGT 3-77 (AC 65-9A)
If a voltage source is attached to a semi-conductor diode with a positive terminal connected to the P material, and a negative terminal to the N material, it is said to be forward-biased and will conduct.

8080. Answer A. JSGT 3-78 (AC 65-9A)
An open at R_1 removes the ground from the base of the N-P-N transistor. This results in zero bias in the base-emitter. Therefore, with zero bias no current can flow through the transistor and the light cannot illuminate.

8081. Answer A. JSGT 3-78 (AC 65-9A)
If R_2 sticks in the up position, maximum current flows to the base of the N-P-N transistor. This results in maximum bias of the base-emitter. With maximum bias in the base-emitter and reverse bias in the base-collector, maximum current flows to the light and it illuminates fully.

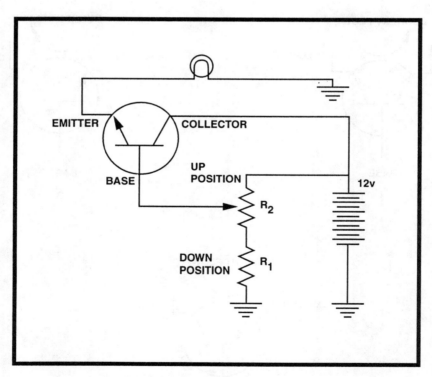

FIGURE 23.—Transistorized Circuit.

8082.   A05
(Refer to figure 24) Which statement concerning the depicted logic gate is true?

A — Any input being 1 will produce a 0 output.
B — Any input being 1 will produce a 1 output.
C — All inputs must be 1 to produce a 1 output.

8082. Answer B. JSGT 3-87 (FA-150-1)
Figure 24 represents a logic OR gate. In an OR gate, any input of 1 (on) results in an output of 1 (on) (answer B). For example, if input number 1, OR input number 2, OR input number 3 are 1 (on), the output will be 1 (on).

Basic Electricity

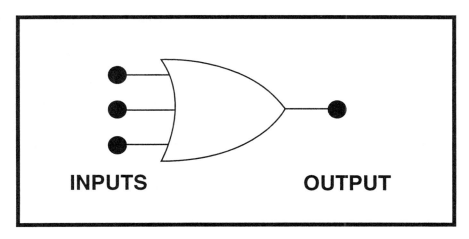

FIGURE 24.—Logic Gate.

8083. A05
(Refer to figure 25) In a functional and operating circuit, the depicted logic gate's output will be 0

A — only when all inputs are 0.
B — when all inputs are 1.
C — when one or more inputs are 0.

8083. Answer C. JSGT 3-87 (FA-150-1)
Figure 25 represents a logic AND gate. In an AND gate, every input must be 1 (on) in order for the output to be 1 (on). For example, input number 1, AND input number 2, AND input number 3 must be 1 (on) for the output to be 1 (on). However, if one or more inputs are 0 (off) the output will be 0 (off) (answer C). Answer (A) is incorrect because having all inputs be 0 is not the only condition that results in a 0 output.

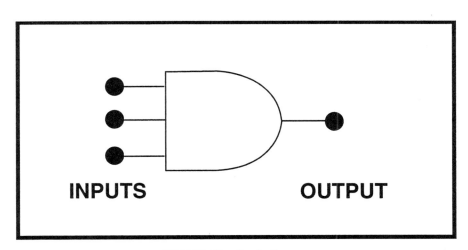

FIGURE 25.—Logic Gate.

8084.   **A05**
(Refer to figure 26) Which of the logic gate output conditions is correct with respect to the given inputs?

A — 1.
B — 2.
C — 3.

8084. Answer B. JSGT 3-87 (FA-150-1)
The illustrations in figure 26 represent exclusive OR gates. This type of logic gate is designed to produce a 1 (on) output whenever the two inputs are dissimilar. Selection 2 (answer B) is the only illustration that has an output of 1 (on) with two dissimilar inputs.

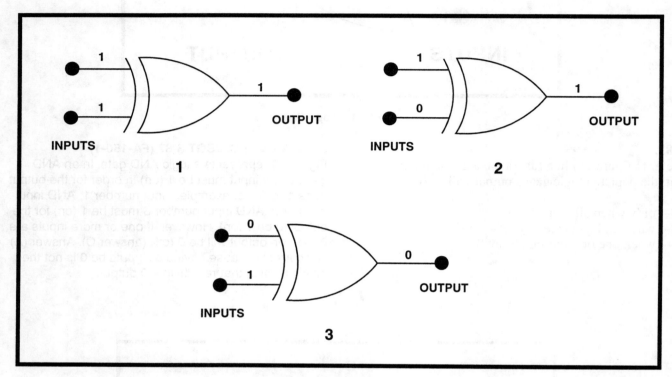

FIGURE 26.—Logic Gates.

SECTION F
ELECTRICAL MEASURING INSTRUMENTS

This section introduces the various types of electrical measuring devices the A&P technician could encounter. While this section presents valuable information on ohmmeters, ammeters, milliammeters, and microammeters, there are no FAA Test questions on this section.

SECTION G
CIRCUIT ANALYSIS

Section G of Chapter 3 draws on the information presented in the previous sections to introduce the principles of troubleshooting. In addition to basic system and component troubleshooting, Section G discusses some common aircraft electrical circuits and explains their basic operating principles. The following FAA Test questions are drawn from this section:

    8057, 8058, 8059, 8060, 8061, 8062, 8063, 8064, 8067, 8068, 8069, 8070, 8071, 8072, 8073.

Basic Electricity

8057. A05
(Refer to figure 15) With the landing gear retracted, the red indicator light will not come on if an open occurs in wire

A — number 19.
B — number 7.
C — number 17.

8058. A05
(Refer to figure 15) The No. 7 wire is used to

A — complete the PUSH-TO-TEST circuit.
B — open the UP indicator light circuit when the landing gear is retracted.
C — close the UP indicator light circuit when the landing gear is retracted.

8059. A05
(Refer to figure 15) When the landing gear is down, the green light will not come on if an open occurs in wire

A — number 7.
B — number 6.
C — number 17.

8057. Answer A. JSGT 3-103 (AC 65-9A)
With the up limit switch in the gear up position, power is supplied to the red light from the bus through the 5 amp breaker, wire #19 and then wire #8. The red indicator light will not come on if a break occurs in either wire #19 or #8. Wire #7 (answer B) and #17 (answer C) supply current to the press-to-test circuit for the red and green lights, respectively.

8058. Answer A. JSGT 3-103 (AC 65-9A)
Wire #7 supplies power to the #18 and #17 wires of the press-to-test function on both the red and green indicator lights. This system allows the flight crew to make certain that the bulb is not burned out.

8059. Answer B. JSGT 3-103 (AC 65-9A)
When the landing gear is in the down position, power is supplied to the green light from the bus through the 5 amp breaker, then wire #6 through the nose gear down switch, then wire numbers #5, #4, and #3. A break in any of these would prevent the light from illuminating. Wire #7 (answer A) and #17 (answer C) supply current to the press-to-test circuit.

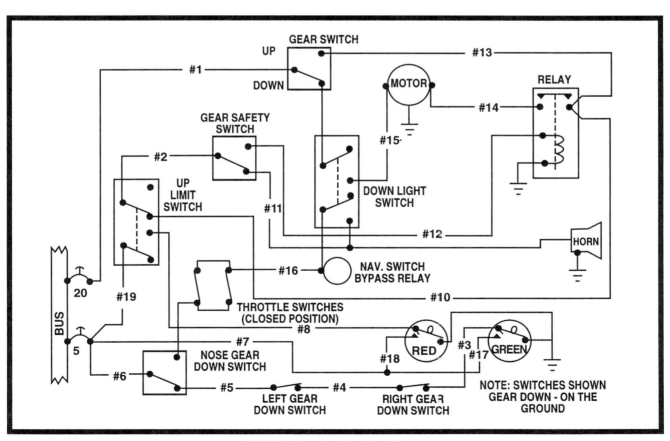

FIGURE 15.—Landing Gear Circuit.

8060.   A05
(Refer to figure 16) What will be the effect if the PCO relay fails to operate when the left-hand tank is selected?

A — The fuel pressure crossfeed valve will not open.
B — The fuel tank crossfeed valve open light will illuminate.
C — The fuel pressure crossfeed valve open light will not illuminate.

8060. Answer C. JSGT 3-104 (AC 65-9A)
If the PCO relay does not operate, switch 13 will not be able to close. Switch 13 and switch 15 must both be closed to supply power to the fuel pressure crossfeed valve open light in the cockpit. With switch 13 open, the light will not illuminate.

8061.   A05
(Refer to figure 16) The TCO relay will operate if 24-volts dc is applied to the bus and the fuel tank selector is in the

A — right-hand tank position.
B — crossfeed position.
C — left-hand tank position.

8061. Answer B. JSGT 3-104 (AC 65-9A)
When the fuel selector switch is in the crossfeed position, power is supplied to the FCF relay, which in turn powers switch number 17. Through this switch the crossfeed valve is energized, closing switch 19 and allowing relay TCO to be energized.

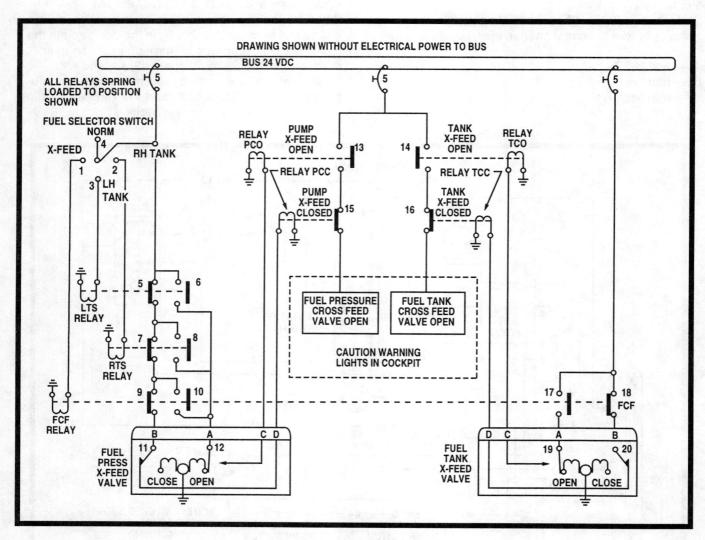

FIGURE 16.—Fuel System Circuit.

Basic Electricity

8062. A05
(Refer to figure 16) With power to the bus and the fuel selector switched to the right-hand tank, how many relays in the system are operating?

A — Three.
B — Two.
C — Four.

8063. A05
(Refer to figure 16) When electrical power is applied to the bus, which relays are energized?

A — PCC and TCC.
B — TCC and TCO.
C — PCO and PCC.

8064. A05
(Refer to figure 16) Energize the circuit with the fuel tank selector switch selected to the left-hand position. Using the schematic, identify the switches that will change position.

A — 5, 9, 10, 11, 12, 13, 15.
B — 3, 5, 6, 7, 11, 13.
C — 5, 6, 11, 12, 13, 15, 16.

8062. Answer A. JSGT 3-104 (AC 65-9A)
When the system has power to the bus, and the fuel selector is switched to the right-hand tank, power is fed from the bus to the RTS relay. This relay opens switch 7 and closes switch 8. Opening switch 7 removes power from cross-feed valve switch 11 which in turn removes power from relay PCC causing switch 15 to open. When switch 8 closes, power flows to switch 12 in the cross-feed valve and feeds power to relay PCO which closes switch 13. A total of three relays have been operated.

8063. Answer A. JSGT 3-104 (AC 65-9A)
A note at the top left of the schematic tells you that all relays are spring loaded to the position shown. When power is supplied to the bus it has a path through switches 5, 7, 9, and 11 to the PCC relay. Power also has a path through switches 18 and 20 to relay TCC.

8064. Answer C. JSGT 3-104 (AC 65-9A)
With the bus energized and the fuel selector in the left-hand position, relay LTS receives power which changes the position of switches 5 and 6. Opening switch 5 causes switch 11 to open and remove power from relay PCC, allowing switch 15 to close. (Note: switch 15 is shown closed because it is spring loaded to that position when the circuit is not energized.) Closing switch 6 energizes switch 12 which allows power to flow to relay PCO and close switch 13. On the other side of the circuit, when power is supplied to the bus, relay TCC energizes and opens switch 16.

8067. A05
(Refer to figure 18) When the landing gears are up and the throttles are retarded, the warning horn will not sound if an open occurs in wire

A — No. 4.
B — No. 2.
C — No. 9.

8068. A05
(Refer to figure 18) The control valve switch must be placed in the neutral position when the landing gears are down to

A — permit the test circuit to operate.
B — prevent the warning horn from sounding when the throttles are closed.
C — remove the ground from the green light.

8069. A05
(Refer to figure 19) Under which condition will a ground be provided for the warning horn through both gear switches when the throttles are closed?

A — Right gear up and left gear down.
B — Both gears up and the control valve out of neutral.
C — Left gear up and right gear down.

8067. Answer A. JSGT 3-102 (AC 65-9A)
The warning horn receives power from the bus through wire #7. For the horn to sound, the circuit must be completed from the horn to ground through wire #6, the throttle switch (which is closed when the throttles are retarded), wire #4, the left gear switch (drawn in the down position), and finally wire #14 to ground. If wire #4 were to break, the circuit could not be completed.

8068. Answer B. JSGT 3-102 (AC 65-9A)
When the gear is down, you do not want the warning horn to sound when you retard the throttles. If the control valve switch were not in the neutral position, the warning horn would have a path to ground through wires #6, #5, #10, #11, #3, and #14.

8069. Answer C. JSGT 3-102 (AC 65-9A)
The only way the warning horn can be grounded through both gear switches is if the left gear is up and the right gear is down. Trace the circuit from the 28V source through wire #10 to the horn, then wire #11 to the throttle switches, which are closed in this problem. After current passes through the throttle switches, it continues through wire #12 to the left gear switch which must be in the up position to provide a path through wire #5 to the right gear switch which must be down to complete the circuit.

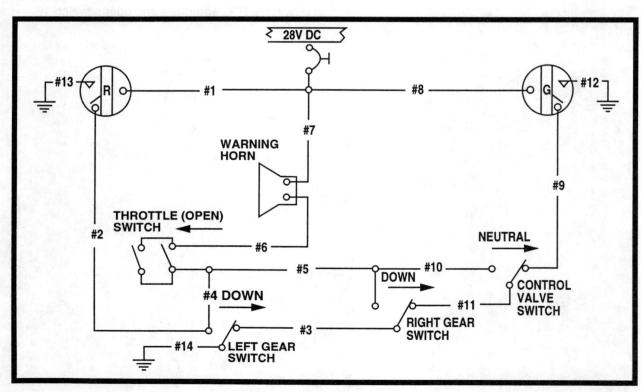

FIGURE 18.—Landing Gear Circuit.

Basic Electricity

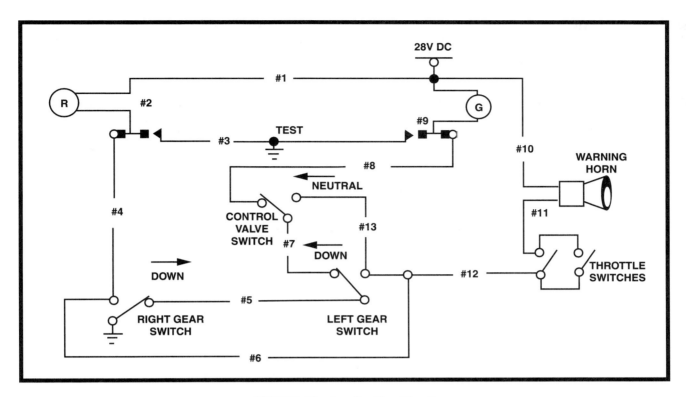

FIGURE 19.—Landing Gear Circuit.

8070.   A05
(Refer to figure 19) When the throttles are retarded with only the right gear down, the warning horn will not sound if an open occurs in wire

A — No. 5.
B — No. 13.
C — No. 6.

8070. Answer A. JSGT 3-102 (AC 65-9A)
With the conditions described, trace the circuit from the 28V source through wire #10 to the horn, then wire #11 to the throttle switches, which are closed in this problem. After current passes through the throttle switches, it passes through wire #12 to the left gear switch which must be in the up position to provide a path through wire #5 to the right gear switch which must be down to complete the circuit. If a break in wire #5 occurs, the warning horn will not sound.

A.8.0.7.1.A.1   A05
(Refer to Figure 19.) When the landing gears are up and the throttles are retarded, the warning horn will not sound if an open occurs in wire

A — No. 6.
B — No. 5.
C — No. 7.

8071. Answer A. JSGT 3-102 (AC 65-9A)
Under the conditions described, trace the circuit from the source through wire #10 to the warning horn, then wire #11 to the throttle switches, which would be closed. After current passes through the throttle switches, it continues through wire #12, wire #6, and the grounded right gear switch, which is in the up position. If wire #6 were open, the warning horn would not sound.

8072.   A05
When referring to an electrical circuit diagram, what point is considered to be at zero voltage?

A — The circuit breaker.
B — The fuse.
C — The ground reference.

8072. Answer C. JSGT 3-100 (AC 65-9A)
The common reference point in a circuit is called the ground. This is the reference point from which most circuit voltages are measured, and is normally considered to be at zero potential.

8073.   A05
(Refer to figure 20) Troubleshooting an open circuit with a voltmeter as shown in this circuit will

A — permit current to flow and illuminate the lamp.
B — create a low resistance path and the current flow will be greater than normal.
C — permit the battery voltage to appear on the voltmeter.

8073. Answer C. JSGT 3-100 (AC 65-9A)
When the voltmeter is connected across the open resistor, the voltmeter closes the circuit by paralleling (shunting) the burned-out resistor. This allows current to flow from the negative terminal of the battery, through the switch, through the voltmeter and lamp, and back to the positive terminal of the battery. Since the resistance of the voltmeter is so high only a small amount of current flows in the circuit. The current is too low to light the lamp, but the voltmeter will read the battery voltage.

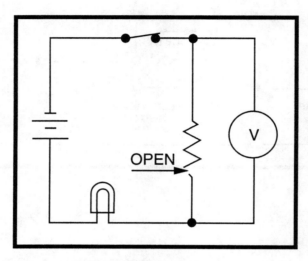

FIGURE 20.—Circuit Diagram.

CHAPTER 4
ELECTRICAL GENERATORS AND MOTORS

Chapter 4 in your textbook discusses several aspects of electrical generators and motors. By studying this chapter, you will become familiar with the various components associated with DC generators, alternators, and motors. In addition, information describing how to test and service these components is presented. While the material presented is valuable to the A&P technician, there are no FAA questions covered in this chapter.

AIRCRAFT DRAWINGS

SECTION A
TYPES OF DRAWINGS

Section A of Chapter 5 begins by examining the three different types of working drawings: the detail drawing, assembly drawing, and installation drawing. The text continues by presenting the common methods of illustrating objects and looks at examples of each. The FAA Test questions pertaining to this section include:

8105, 8106, 8108, 8109, 8111, 8112, 8116, 8119, 8131, 8134, 8136, 8137, 8138, 8139, 8140, 8512, 8513, 8514.

8105.   B01
(1) A detail drawing is a description of a single part.
(2) An assembly drawing is a description of an object made up of two or more parts.
Regarding the above statements,

A — only No. 1 is true.
B — neither No. 1 nor No. 2 is true.
C — both No. 1 and No. 2 are true.

8105. Answer C. JSGT 5-2 (AC 65-9A)
Both statements 1 and 2 are correct. Detail drawings depict a single part, and usually give information about its size, shape, material, and method of manufacture. Assembly drawings, on the other hand, depict an object made up of two or more parts.

8106.   **B01**
(Refer to figure 28) Identify the bottom view of the object shown.

A — 1.
B — 2.
C — 3.

8106. Answer B. JSGT 5-10 (AC 65-9A)
When you rotate the front view 90 degrees to obtain a bottom view, the entire bottom surface is visible and is therefore depicted by four solid outlines. The channel in the center of the part is not visible so its sides would be projected by the use of hidden (dashed) lines, as depicted in selection 2 (answer B). Views 1 and 3 (answers A and C) each have solid lines depicting parts of the channel that would not be visible from the bottom view.

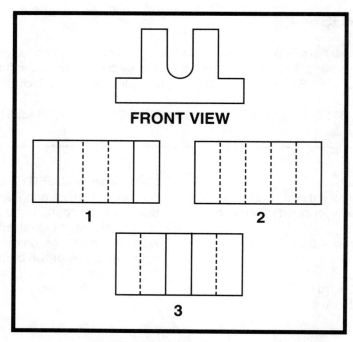

FIGURE 28.—Object Views.

8108.   **B01**
Which statement is true regarding an orthographic projection?

A — There are always at least two views.
B — It could have as many as eight views.
C — One-view, two-view, and three-view drawings are the most common.

8108. Answer C. JSGT 5-10 (AC 65-9A)
When using orthographic projection, as many as six views can be depicted; however, one-, two-, or three-view drawings are most common.

8109.   **B01**
(Refer to figure 29) Identify the left side view of the object shown.

A — 1.
B — 2.
C — 3.

8109. Answer C. JSGT 5-10 (AC 65-9A)
This is an application of orthographic projection. The left side view is obtained by rotating the object 90 degrees, so the left side of the object faces you. Since the entire left side is visible, it is depicted by four solid (visible) lines. The step, on the other hand, is not visible and is depicted by a dashed (hidden) line.

Aircraft Drawings

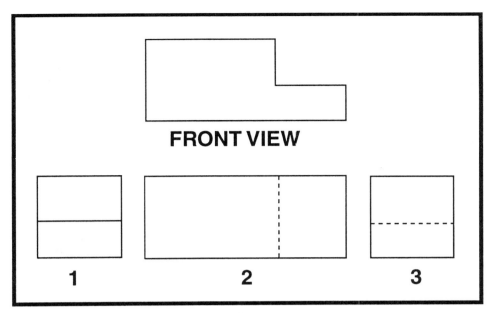

FIGURE 29.—Object Views.

8111.　　B01
(Refer to figure 30) Identify the bottom view of the object.

A — 1.
B — 2.
C — 3.

8111. Answer A. JSGT 5-10 (AC 65-9A)
You must apply the rules of orthographic projection to obtain a bottom view. Once the object is rotated 90 degrees to obtain the bottom view, the four sides that make up the bottom would be visible and, therefore, be depicted with visible (solid) lines. The two steps would not be visible and are represented by hidden (dashed) lines, running vertically at the same distance from the ends that they appear in the front view.

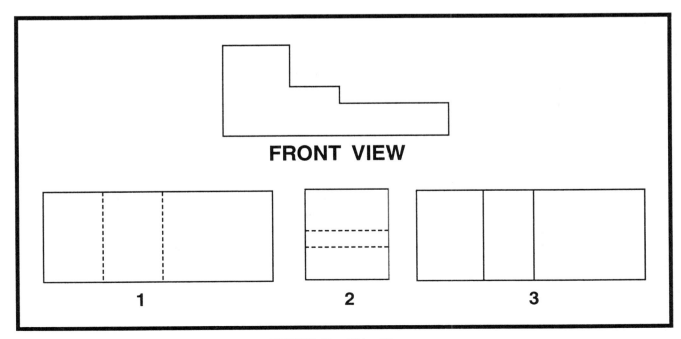

FIGURE 30.—Object Views.

8112. B01
(1) Schematic diagrams indicate the location of individual components in the aircraft.
(2) Schematic diagrams indicate the location of components with respect to each other within the system.
Regarding the above statements,

A — only No. 1 is true.
B — both No. 1 and No. 2 are true.
C — only No. 2 is true.

8116. B02
For sketching purposes, almost all objects are composed of one or some combination of six basic shapes; these include the

A — angle, arc, line, plane, square, and circle.
B — triangle, circle, cube, cylinder, cone, and sphere.
C — triangle, plane, arc, line, square, and polygon.

8119. B02
Working drawings may be divided into three classes. They are:

A — title drawings, installation drawings, and assembly drawings.
B — detail drawings, assembly drawings, and installation drawings.
C — orthographic projection drawings, pictorial drawings, and detail drawings.

8131. B03
One purpose for schematic diagrams is to show the

A — functional location of components within a system.
B — physical location of components within a system.
C — size and shape of components within a system.

8134. B03
A hydraulic system schematic drawing would indicate the

A — specific location of the individual components within the aircraft.
B — direction of fluid flow through the system.
C — type and quantity of the hydraulic fluid.

8136. B03
(1) A measurement should not be scaled from an aircraft print because the paper shrinks or stretches when the print is made.
(2) When a detail drawing is made, it is carefully and accurately drawn to scale, and is dimensioned.
Regarding the above statements,

A — only No. 2 is true.
B — both No. 1 and No. 2 are true.
C — neither No. 1 nor No. 2 is true.

8112. Answer C. JSGT 5-8 (AC 65-9A)
Only statement number 2 is correct. Schematic diagrams are used to explain a principle of operation, rather than show the parts as they actually appear. They do not indicate the location of individual components in the aircraft, but do indicate the location of components with respect to each other within the system.

8116. Answer B. JSGT 5-18 (AC 65-9A)
Almost all objects are comprised of some or a combination of six basic shapes. These include the triangle, circle, cube, cylinder, cone and sphere. Answers (A) and (C) are incorrect because the angle, arc, line, and plane are not shapes.

8119. Answer B. JSGT 5-2 (AC 65-9A)
Working drawings must provide all of the information essential to making and assembling a particular object. They may be divided into three classes: (1) detail drawings, (2) assembly drawings, and (3) installation drawings.

8131. Answer A. JSGT 5-8 (AC 65-9A)
Schematic diagrams illustrate the functional location of components with respect to each other within the system and are used mainly in trouble-shooting.

8134. Answer B. JSGT 5-8 (AC 65-9A)
A schematic diagram is used to explain a principle of operation, rather than to show the parts as they actually appear, or as they function. In the case of hydraulic, fuel, or oil systems, a schematic drawing would show the direction of fluid flow.

8136. Answer B. JSGT 5-2 (AC 65-9A)
These are both true statements. While aircraft drawings are carefully and accurately drawn to scale, various copying processes, moisture, and age may cause the image on a drawing to differ significantly from the size to which it was originally drawn. For this reason, you should not scale off an aircraft drawing.

8137.   B03
The drawings often used in illustrated parts manuals are

A — exploded view drawings.
B — block drawings.
C — detail drawings.

8138.   B03
A drawing in which the subassemblies or parts are shown as brought together on the aircraft is called

A — a sectional drawing.
B — a detail drawing.
C — an installation drawing.

8139.   B03
What type of diagram shows the wire size required for a particular installation?

A — A block diagram.
B — A schematic diagram.
C — A wiring diagram.

8140.   B03
What type of diagram is used to explain a principle of operation, rather than show the parts as they actually appear?

A — A pictorial diagram.
B — A schematic diagram.
C — A block diagram.

8512.   K01
(Refer to figure 62, 62A, & 62B as necessary) Which doubler(s) require(s) heat treatment before installation?

A — -101.
B — -102.
C — Both.

8513.   K01
(Refer to figure 62, 62A & 62B, as necessary) Using only the information given (when bend allowance, set back, etc., have been calculated) which doubler is it possible to construct and install?

A — -101.
B — -102.
C — Both.

8514.   K01
(Refer to figure 62) The -100 in the title block (Area 1) is applicable to which doubler part number(s)?

A — -101.
B — -102.
C — Both.

8137. Answer A. JSGT 5-5 (AC 65-9A)
Illustrated parts lists often use exploded-view drawings to show every part that is in an assembly. All of the parts are shown in their relative positions, but are expanded outward, so that each part can be identified.

8138. Answer C. JSGT 5-3 (AC 65-9A)
All subassemblies or parts are brought together in an installation drawing. The bills of material on these drawings list every fastener needed, and notes on the face of the drawing furnish any information required for the installation.

8139. Answer C. JSGT 5-8 (AC 65-9A)
Electrical wiring diagrams specify the size of wire to be used, the type of terminals, and identify each component. These drawings are included in the aircraft service manual.

8140. Answer B. JSGT 5-8 (AC 65-9A)
A schematic diagram is used to explain a principle of operation, rather than the parts as they actually appear or function.

8512. Answer B. JSGT 5-2 (AC 65-9A)
Areas 2 and 3 in figure 62A supply information on the preparation of these parts. Note that step 3 in area 3 instructs you to heat treat the -102 part to a hardness of -T6. Area 2 gives no heat treat instructions for a -101 part. Answer (B) is correct.

8513. Answer A. JSGT 5-2 (AC 65-9A)
The key to answering this question is in area three — General Notes -200. The -102 calls for process specifications not given in this data. Therefore, it is only possible to construct and install the -101.

8514. Answer A. JSGT 5-2 (Part 43, Appendix A)
Find the -100 which is set vertically in the title block. The second block above it indicates that one -101 doubler is required for a -100 installation. Note that the -200 requires one -102 doubler. Answer (A) is correct.

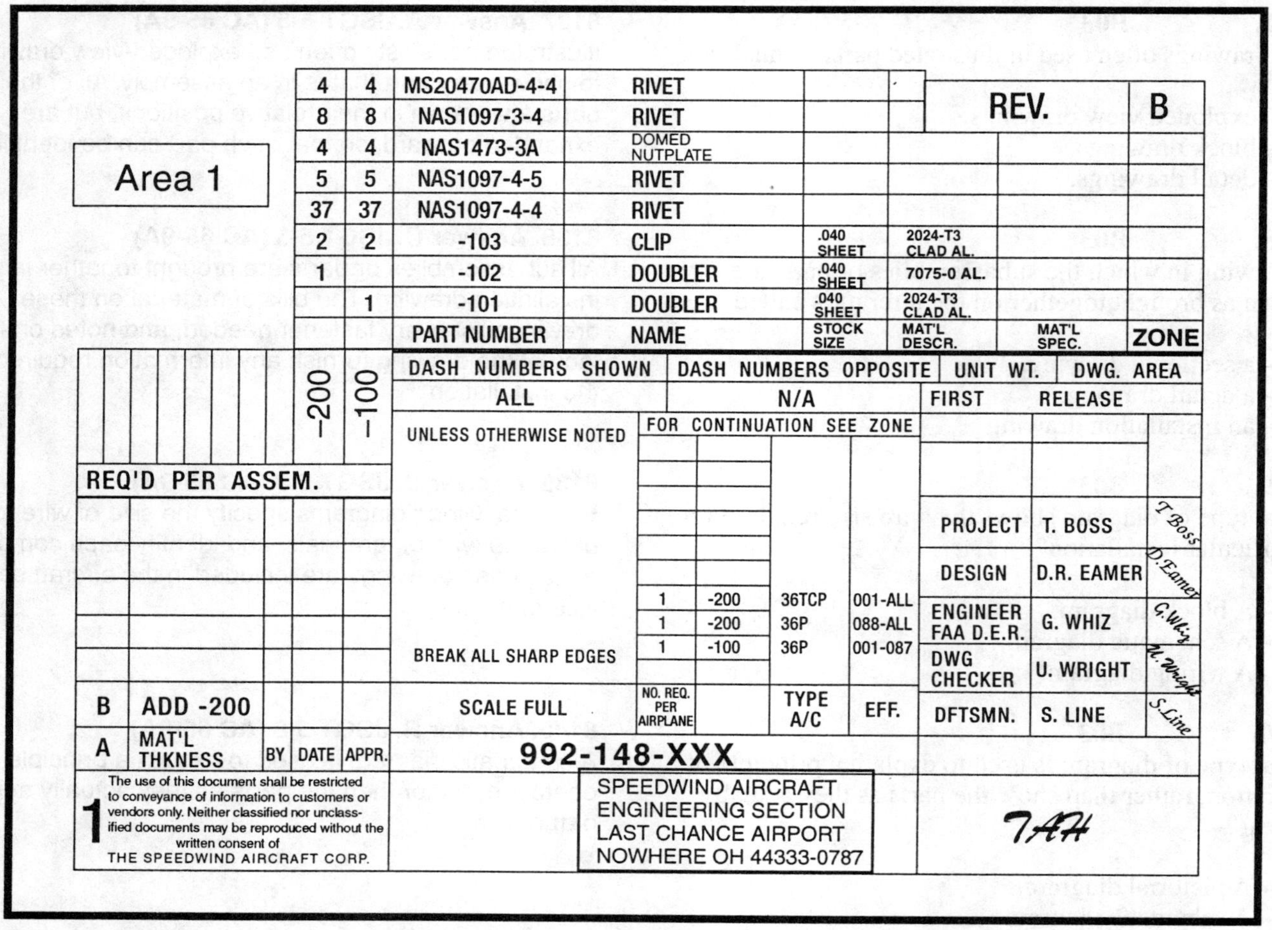

4	4	MS20470AD-4-4	RIVET			REV.	B
8	8	NAS1097-3-4	RIVET				
4	4	NAS1473-3A	DOMED NUTPLATE				
5	5	NAS1097-4-5	RIVET				
37	37	NAS1097-4-4	RIVET				
2	2	-103	CLIP	.040 SHEET	2024-T3 CLAD AL.		
	1	-102	DOUBLER	.040 SHEET	7075-0 AL.		
1		-101	DOUBLER	.040 SHEET	2024-T3 CLAD AL.		
		PART NUMBER	NAME	STOCK SIZE	MAT'L DESCR.	MAT'L SPEC.	ZONE
-200	-100	DASH NUMBERS SHOWN	DASH NUMBERS OPPOSITE	UNIT WT.	DWG. AREA		

Area 1

REQ'D PER ASSEM.

ALL	N/A
UNLESS OTHERWISE NOTED	FOR CONTINUATION SEE ZONE

FIRST RELEASE

			PROJECT — T. BOSS — *T. Boss*
			DESIGN — D.R. EAMER — *D. Eamer*
1	-200	36TCP	001-ALL
1	-200	36P	088-ALL
1	-100	36P	001-087

ENGINEER FAA D.E.R. — G. WHIZ — *G. Whiz*
DWG CHECKER — U. WRIGHT — *U. Wright*

BREAK ALL SHARP EDGES

SCALE FULL	NO. REQ. PER AIRPLANE	TYPE A/C	EFF.	DFTSMN. — S. LINE — *S. Line*

B	ADD -200
A	MAT'L THKNESS

BY | DATE | APPR.

992-148-XXX

SPEEDWIND AIRCRAFT
ENGINEERING SECTION
LAST CHANCE AIRPORT
NOWHERE OH 44333-0787

TAH

1 — The use of this document shall be restricted to conveyance of information to customers or vendors only. Neither classified nor unclassified documents may be reproduced without the written consent of THE SPEEDWIND AIRCRAFT CORP.

FIGURE 62.—Part 1 of 3 — Maintenance Data.

Aircraft Drawings

> Area 2

GENERAL NOTES – 100

1. All bends +/– .5°
2. All holes +/– .003.
3. Apply Alodine 1000.
4. Prime with MIL-P-23377 or equivalent.
5. Trim S-1 C just aft of the clip at STA. 355.750 and forward of the front face of the STA. 370.25 frame and remove from the airplane.
6. Position the -101 doubler as shown. Install wet with NAS1097AD-4-4 and -4-5 rivets and a faying surface seal of PR 1422. Pick up the rivet row that was in S-1 C and the aft rivets in STA 370.25. Tie doubler into front frame with clips as shown using MS20470AD-4-4 rivets through the clips and the frame.
7. Install 4 NAS1473-3A nutplates with NAS1097-3-4 rivets through the skin and doubler to retain the antenna.
8. Strip paint and primer from under the antenna footprint.
9. Treat skin with Alodine 1000.
10. Install antenna and apply weather seal fillet around antenna base.

> Area 3

GENERAL NOTES – 200

Note: P. S. = Process Specification
IAW = in accordance with

1. All bends IAW P. S. 1000.
2. All holes IAW P. S. 1015.
3. Heat treat -102 to -T6 IAW P. S. 5602.
4. Alodine IAW P. S. 10000.
5. Prime IAW P. S. 10125.
6. Trim S-1 C just aft of the clip at STA. 355.750 and forward of the front face of the STA. 370.25 frame and remove from the airplane.
7. Position the -102 doubler as shown. Install wet with NAS1097AD-4-4 and -4-5 rivets, and a faying surface seal IAW P. S. of 41255. Pick up the rivet row that was in S-1 C and the aft rivets in STA 370.25. Add two edge rows as shown. Tie doubler into front frame with clips as shown using MS20470AD-4-4 rivets through the clips and the frame.
8. Install 4 NAS1473-3A nutplates with NAS1097-3-4 rivets through the skin and doubler to retain the antenna.
9. Strip paint and primer from under the antenna footprint.
10. Treat skin IAW P. S. 10000.
11. Install antenna and apply weather seal fillet around antenna base.

FIGURE 62A.—Part 2 of 3 — Maintenance Data.

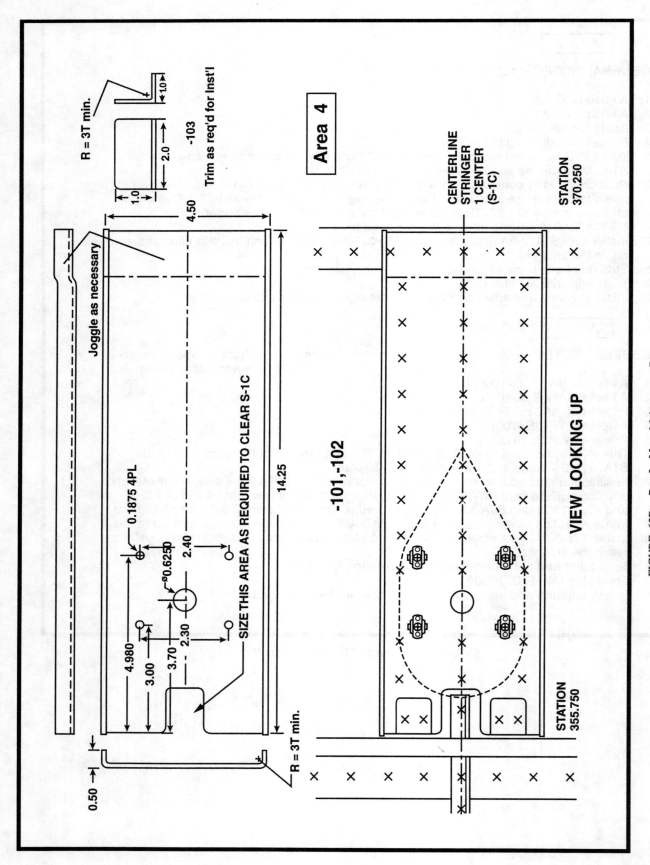

FIGURE 62B.—Part 3 of 3 — Maintenance Data.

SECTION B
DRAWING PRACTICES

This section describes and presents accepted industry drawing practices. Included is a discussion on the various types of lines used in aircraft drawings as well as the symbology used to represent different materials. Another important topic within this section deals with the steps and procedures used to construct a sketch. The following FAA Test questions are included in this section:

8103, 8104, 8107, 8110, 8113, 8114, 8115, 8117, 8118, 8120, 8121, 8122, 8123, 8124, 8125, 8126, 8127, 8128, 8129, 8130, 8132, 8133, 8135, 8141.

8103.　　　B01
What type of line is normally used in a mechanical drawing or blueprint to represent an edge or object not visible to the viewer?

A — Medium-weight dashed line.
B — Medium solid line.
C — Alternate short and long light dashes.

8103. Answer A. JSGT 5-13 (AC 65-9A)
The line described in this question is called a hidden line and is represented by a medium-weight dashed line.

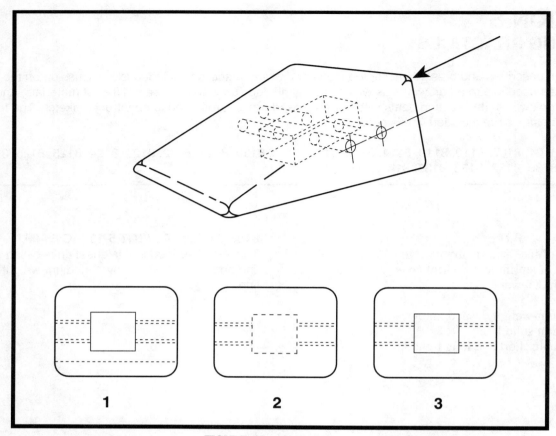

FIGURE 27.—Object Views.

8104.　　　B01
(Refer to figure 27) In the isometric view of a typical aileron balance weight, identify the view indicated by the arrow.

A — 1.
B — 3.
C — 2.

8104. Answer B. JSGT 5-13 (AC 65-9A)
The first thing you must realize is that the back side of the weight is open to the internal cavity. This means that when viewed from the back, the opening will be represented by a solid line. The four openings coming in from the sides are not visible when viewed from the back, so they appear as hidden lines (dashes). Because the side openings are not on the same center line, they appear offset when viewed from the rear. The rounded nose of the weight has no edges or corners, so it is not depicted in the rear view.

8107.　　　B01
A specific measured distance from the datum or some other point identified by the manufacturer, to a point in or on the aircraft is called a

A — zone number.
B — specification number.
C — station number.

8107. Answer C. JSGT 5-17 (AC 65-9A)
Manufacturers use station numbers to identify the location of items on an aircraft. Station numbers is typically measured in inches forword or aft of a datum or some other point identified by the manufacturer..

8110.　　　B01
A line used to show an edge which is not visible is a

A — phantom line.
B — hidden line.
C — break line.

8110. Answer B. JSGT 5-13 (AC 65-9A)
A medium-weight dashed line, known as a hidden line, is used to indicate edges, or parts, which are not visible.

8113.   B02
(Refer to figure 31) What are the proper procedural steps for sketching repairs and alterations?

A — 3, 1, 4, 2.
B — 4, 2, 3, 1.
C — 1, 3, 4, 2.

8114.   B02
Which statement is applicable when using a sketch for making a part?

A — The sketch may be used only if supplemented with three-view orthographic projection drawings.
B — The sketch must show all information to manufacture the part.
C — The sketch need not show all necessary construction details.

8113. Answer A. JSGT 5-10 (AC 65-9A)
There are four basic steps in making a sketch. First, determine what views are necessary to portray the object and block in the views using light construction lines. Second, complete the details, and darken the object outline. Third, sketch extension and dimension lines, and add detail. Finally, complete the drawing by adding notes, dimensions, a title, and a date.

8114. Answer B. JSGT 5-18 (AC 65-9A)
A sketch may be used to manufacture a replacement part, when necessary. The sketch must provide all of the necessary information to fabricate the part.

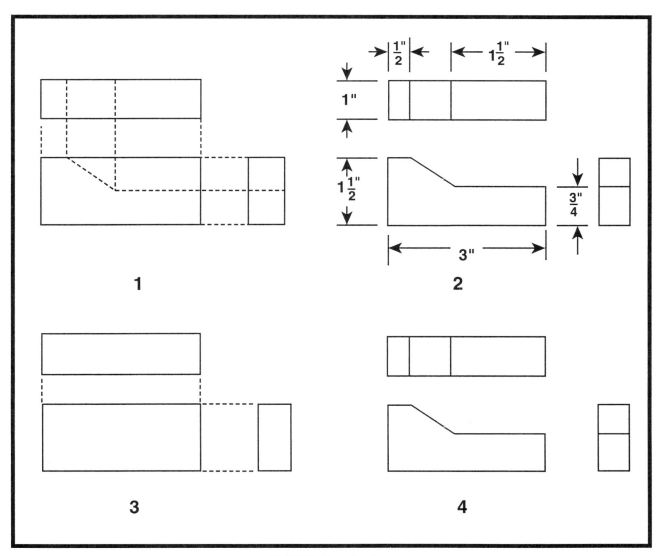

FIGURE 31.—Sketches.

8115. **B02**
(Refer to figure 32) What is the next step required for a working sketch of the illustration?

A — Darken the object outlines.
B — Sketch extension and dimension lines.
C — Add notes, dimensions, title, and date.

8117. **B02**
What should be the first step of the procedure in sketching an aircraft wing skin repair?

A — Draw heavy guidelines.
B — Lay out the repair.
C — Block in the views.

8115. Answer B. JSGT 5-18 (AC 65-9A)
When making a sketch, after the details have been added and the object lines darkened, the next step is to sketch the extension and dimension lines.

8117. Answer C. JSGT 5-18 (AC 65-9A)
There are four basic steps in making a sketch. First, determine what views are necessary to portray the object and block in the views using light construction lines. Second, complete the details and darken the object outline. Third, sketch extension and dimension lines and add detail. Finally, complete the drawing by adding notes, dimensions, a title, and a date.

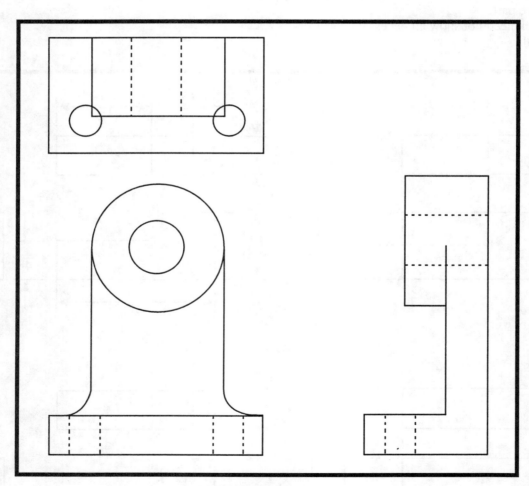

FIGURE 32.—Sketches.

8118. B02
(1) According to FAR Part 91, repairs to an aircraft skin should have a detailed dimensional sketch included in the permanent records.
(2) On occasion, a mechanic may need to make a simple sketch of a proposed repair to an aircraft, a new design, or a modification.
Regarding the above statements,

A — only No. 1 is true.
B — only No. 2 is true.
C — Both No. 1 and No. 2 are true.

8120. B02
Since sketches, by their nature, are drawn without the use of drafting instruments, the layout process is usually made easier by the use of

A — graph paper.
B — plain white paper.
C — ruled note paper.

8121. B02
What material symbol is frequently used in drawings to represent all metals?

A — Steel.
B — Cast iron.
C — Aluminum.

8122. B02
(Refer to figure 33) Which material section-line symbol indicates cast iron?

A — 1.
B — 2.
C — 3.

8118. Answer B. JSGT 5-18 (AC 65-9A)
Only statement number 2 is correct. In executing FAA Form 337, the mechanic may need to make a sketch to describe a repair or alteration. The amount of detail necessary is only that which is required to adequately describe the work accomplished. Statement (1) is false. FAR Part 91 only requires a description or reference to acceptable data, when work is performed. There is no requirement for a detailed dimension sketch to be included in an aircraft's permanent records.

8120. Answer A. JSGT 5-18 (AC 65-9A)
When making a sketch, it is typically best to use a pencil and graph paper. The graph paper allows you to more accurately sketch to scale as well as maintain the proper perspective. Answers (B) and (C) are incorrect because most people find it more difficult to sketch objects on either plain white or ruled paper.

8121. Answer B. JSGT 5-13 (AC 65-9A)
At times, a material may not be indicated symbolically when its exact specification must be shown elsewhere on the drawing. In this case, the diagonal section lines representing cast iron are often used as a generic section line, and the material specification is listed in the bill of materials or a note.

8122. Answer C. JSGT 5-13 (AC 65-9A)
Cast iron is indicated by thin parallel lines drawn on a 45° angle to the bottom of the drawing. This type of sectioning line is used for other materials when the exact specification for the material used appears elsewhere on the drawing.

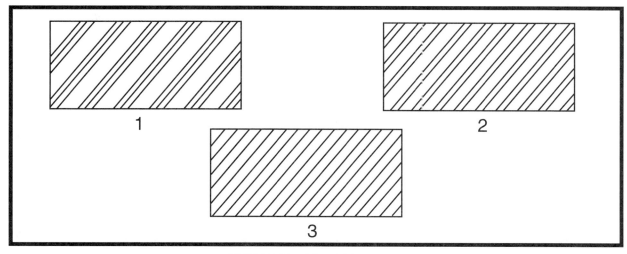

FIGURE 33.—Material Symbols.

8123.   **B03**
(Refer to figure 34) What is the dimension of the chamfer?

A — 1/16 × 37°.
B — 0.3125 + .005 −0.
C — 0.0625 × 45°.

8123. Answer C. JSGT 5-12 (AC 65-9A)
A chamfer is the corner of an object that has been tapered (usually at 45°) to relieve stress, ease assembly, or prevent damage to the part. The chamfer and its dimensions shown in figure 34 are depicted on the far left of the top drawing. Both the length and angle of a chamfer are typically given in a detailed drawing. In this question, you must convert the fraction given to a decimal. In this case, 1/16 inch is equivalent to 0.0625. Therefore, the dimension of the chamfer is 0.0625 × 45°.

8124.   **B03**
(Refer to figure 34) What is the maximum diameter of the hole for the clevis pin?

A — 0.3175.
B — 0.3130.
C — 0.31255.

8124. Answer A. JSGT 5-14 (AC 65-9A)
While neither hole on the drawing is labeled as a clevis pin, you should know that a clevis pin goes in a through hole. This eliminates the possibility that the question refers to the blind hole machined into the left half of the drawing. The top view depicts a through hole on the right side of the drawing with a dimension of .3125 (+.005, −.000). The maximum dimension of the hole is .3175 (.3125 + .005 = .3175).

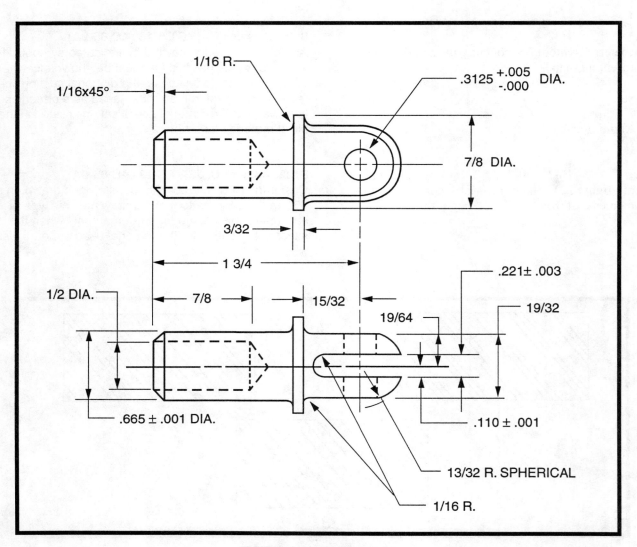

FIGURE 34.—Aircraft Drawing.

8125. **B03**
(Refer to figure 34) What would be the minimum diameter of 4130 round stock required for the construction of the clevis that would produce a machined surface?

A — 55/64 inch.
B — 1 inch.
C — 7/8 inch.

8126. **B03**
(Refer to figure 34) Using the information, what size drill would be required to drill the clevis bolthole?

A — 5/16 inch.
B — 21/64 inch.
C — 1/2 inch.

B.8.1.2.7.B.1 **B03**
The measurements showing the ideal or "perfect" sizes of parts on drawings are called

A — tolerances.
B — allowances.
C — dimensions.

8128. **B03**
(Refer to figure 35) Identify the extension line.

A — 3.
B — 1.
C — 4.

8125. Answer B. JSGT 5-12 (AC 65-9A)
The largest diameter shown on the drawing is 7/8 inch. Since a machined surface is required, the stock material must have a diameter larger than 7/8 inch. The only diameter listed that is greater than 7/8 inch is 1 inch (answer B).

8126. Answer A. JSGT 5-14 (AC 65-9A)
The clevis bolthole diameter is located on the right side of the top drawing. The hole diameter is .3125, which is the decimal equivalent of 5/16.

8127. Answer C JSGT 5-14 (AC 65-9A)
A drawing, to be meaningful, not only must show the shape of the part, but it must accurately give all needed dimensions. This is accomplished with dimensioning.

8128. Answer A. JSGT 5-12 (AC 65-9A)
Extension lines are light lines extending from the point where the measurement is made to a convenient clear area where the dimension may be written without cluttering the drawing. Selection 3 (answer A) identifies an extension line.

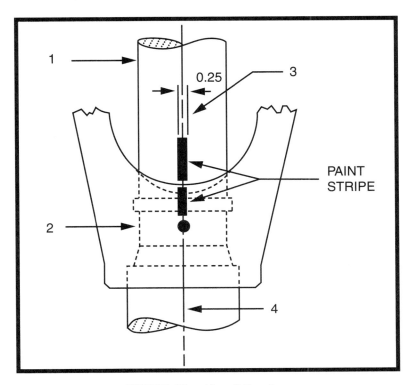

FIGURE 35.—Aircraft Drawing.

8129. B03

(Refer to figure 36) The diameter of the holes in the finished object is

A — 3/4 inch.
B — 31/64 inch.
C — 1/2 inch.

8130. B03

Zone numbers on aircraft blueprints are used to

A — locate parts, sections, and views on large drawings.
B — indicate different sections of the aircraft.
C — locate parts in the aircraft.

8132. B03

When reading a blueprint, a dimension is given as 4.387 inches +.005 -.002. Which statement is true?

A — The maximum acceptable size is 4.390 inches.
B — The minimum acceptable size is 4.385 inches.
C — The minimum acceptable size is 4.382 inches.

8129. Answer C. JSGT 5-12 (AC 65-9A)
Each hole is referenced to Note 1. Note 1 specifies that the holes should be initially drilled slightly undersize with a 31/64 inch drill, and finished with a 1/2 inch ream.

8130. Answer A. JSGT 5-16 (AC 65-9A)
Zone numbers on drawings are similar to the numbers and letters printed on the borders of a map. They are used to help locate a particular point, or part on large drawings.

8132. Answer B. JSGT 5-14 (AC 65-9A)
Given the dimension 4.387 (+.005 –.002), you can add .005 to get a maximum size of 4.392, and subtract .002 to get a minimum size of 4.385.

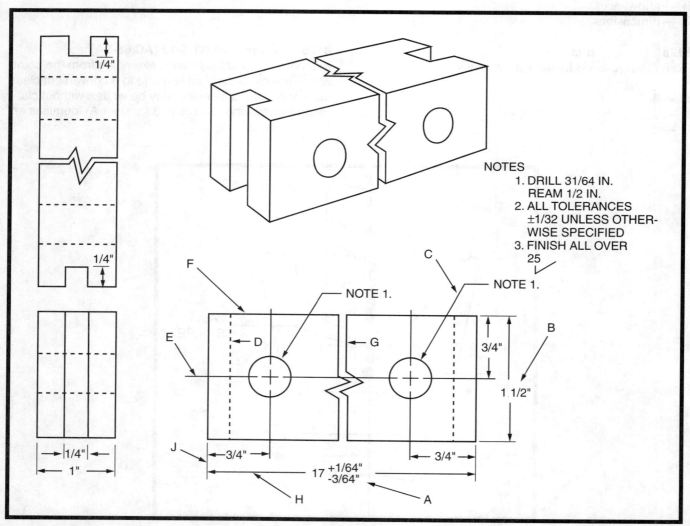

FIGURE 36.—Aircraft Drawing.

Aircraft Drawings

8133.   **B03**
What is the allowable manufacturing tolerance for a bushing where the outside dimensions shown on the blueprint are: 1.0625 +.0025 −.0003?

A — .0028.
B — 1.0650.
C — 1.0647.

8135.   **B03**
(Refer to figure 37) The vertical distance between the top of the plate and the bottom of the lowest 15/64-inch hole is

A — 2.250.
B — 2.242.
C — 2.367.

8141.   **B03**
In the reading of aircraft blueprints, the term "tolerance", used in association with aircraft parts or components,

A — is the tightest permissible fit for proper construction and operation of mating parts.
B — is the difference between extreme permissible dimensions that a part may have and still be acceptable.
C — represents the limit of galvanic compatibility between different adjoining material types in aircraft parts.

8133. Answer A. JSGT 5-14 (AC 65-9A)
Tolerance is the difference between the extreme permissible dimensions. Here, the upper limit is +.0025 and the lower limit is .0003. The difference is .0028 [.0025 − (−.0003) = .0028].

8135. Answer C. JSGT 5-15 (AC 65-9A)
To determine the distance, you must add the incremental distances depicted. The distances which need to be added are as follows: From the top of the plate to the center of the first hole (3/8"); from the center of the first hole to the center of the second hole (7/8"); from the center of the second hole to the center of the third hole (7/8"); from the center of the third hole to the center of the fourth hole (1/8"); and, finally, the distance from the center of the fourth hole to the bottom of the fourth hole (15/128"). Total these distances and convert to the decimal equivalent for a total measurement of 2.367".

8141. Answer B. JSGT 5-14 (AC 65-9A)
Tolerance is the difference between extreme permissible dimensions, or the range of error that will be accepted or tolerated in a serviceable part. On the other hand, allowance is the difference between upper and lower dimensions, and represents the tightest permissible fit. Therefore, answer (B) is correct.

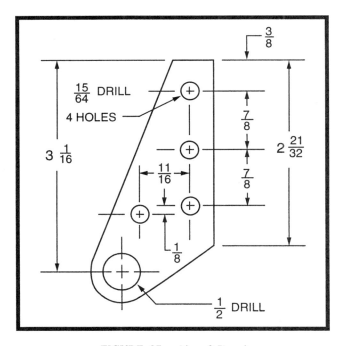

FIGURE 37.—Aircraft Drawing.

SECTION C
CHARTS AND GRAPHS

Section C looks at information that is presented in graphical and chart form. Since charts and graphs are common in the maintenance industry, you must be proficient in their use. This section looks at several charts, including performance, electrical, and cable tension charts. Included in this section are the following FAA Test questions:

    8142, 8143, 8144, 8145, 8146, 8147, 8148, 8149, 8150, 8151, 8152.

8142.   **B04**
(Refer to figure 38) An aircraft reciprocating engine has a 1,830 cubic-inch displacement and develops 1,250 brake-horsepower at 2,500 RPM. What is the brake mean effective pressure?

A — 217.
B — 205.
C — 225.

8142. Answer A. JSGT 5-23 (AC 65-9A)
To find this answer, begin by locating 1,250 HP on the top of the chart. From this value, drop down vertically until you reach the line representing 1,830 cubic inches of displacement. From this intersection, extend a line horizontally to the right until you intercept the line representing 2,500 RPM. Now, drop down vertically to read the brake mean effective pressure on the bottom line of the chart. The brake mean effective pressure is 217.

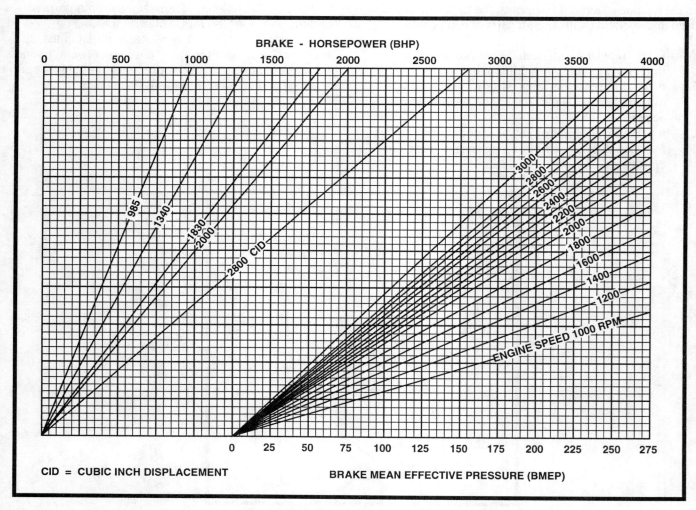

FIGURE 38.—Performance Chart.

8143.   **B04**
(Refer to figure 38) An aircraft reciprocating engine has a 2,800 cubic-inch displacement, develops 2,000 brake-horsepower, and indicates 270 brake mean effective pressure. What is the engine speed (RPM)?

A — 2,200.
B — 2,100.
C — 2,300.

8144.   **B04**
(Refer to figure 38) An aircraft reciprocating engine has a 2,800 cubic-inch displacement and develops 2,000 brake-horsepower at 2,200 RPM. What is the brake mean effective pressure?

A — 257.5.
B — 242.5.
C — 275.0.

8143. Answer B. JSGT 5-23 (AC 65-9A)
Begin this problem by locating 2,000 HP at the top of the chart. Drop down vertically until you intersect the line representing the 2,800 cu. in. displacement. From this point extend a line horizontally to the right. Now, locate the 270 brake mean effective pressure along the bottom of the chart. From this point extend a line up vertically until it intersects the horizontal line drawn earlier. The intersection of these two lines represents the engine speed (RPM). The answer is 2,100 RPM.

8144. Answer A. JSGT 5-23 (AC 65-9A)
At the top of the chart, locate 2,000 HP. From this value drop down vertically until you intersect the line representing the 2,800 cubic-inch displacement. From this point, extend a line horizontally to the right until you intercept the line representing 2,200 RPM. Now drop down vertically to read the brake mean effective pressure on the bottom line of the chart. The brake mean effective pressure is 257.5.

8145.   **B04**

(Refer to figure 39) Determine the cable size of a 40-foot length of single cable in free air, with a continuous rating, running from a bus to the equipment in a 28-volt system with a 15-ampere load and a 1-volt drop.

A — No. 10.
B — No. 11.
C — No. 18.

8146.   **B04**

(Refer to figure 39) Determine the maximum length of a No. 16 cable to be installed from a bus to the equipment in a 28-volt system with a 25-ampere intermittent load and a 1-volt drop.

A — 8 feet.
B — 10 feet.
C — 12 feet.

8145. Answer A. JSGT 5-22 (AC 43.13-1A)
Locate the column on the left side of the chart that represents a circuit voltage of 28 with a 1 volt drop. Find the horizontal line representing 40 feet and follow it to the right until it intersects the diagonal line indicating 15 amps. Since this point is above curve 2, installation as a single cable in free-air is permitted. Now, drop a vertical line to the bottom of the chart. This line falls between wire sizes #10 and #12. Whenever the chart indicates a wire size between two sizes, you must select the larger wire. In this case you would use #10 wire.

8146. Answer A. JSGT 5-22 (AC 43.13-1A)
Begin by finding the wire size on the bottom of the cart. From the #16, move upward to curve 3, which represents an intermittent load. From this point, project a horizontal line to the left until it reaches the column representing 28 volts with a 1 volt drop. The maximum length of a cable that can safely handle these requirements is 8 feet.

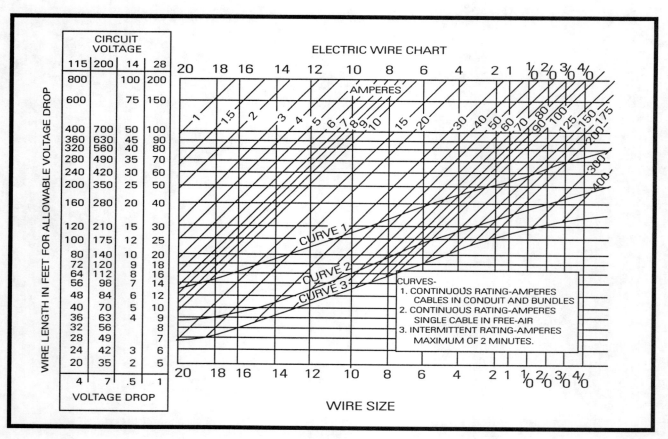

FIGURE 39.—Electric Wire Chart.

Aircraft Drawings

8147. B04
(Refer to figure 39) Determine the minimum wire size of a single cable in a bundle carrying a continuous current of 20 amperes 10 feet from the bus to the equipment in a 28-volt system with an allowable 1-volt drop.

A — No. 12.
B — No. 14.
C — No. 16.

8148. B04
(Refer to figure 39) Determine the maximum length of a No. 12 single cable that can be used between a 28-volt bus and a component utilizing 20 amperes continuous load in free air with a maximum acceptable 1-volt drop.

A — 22.5 feet.
B — 26.5 feet.
C — 12.5 feet.

8149. B04
(Refer to figure 40) Determine the proper tension for a 1/8-inch cable (7 × 19) if the temperature is 80°F.

A — 70 pounds.
B — 75 pounds.
C — 80 pounds.

8147. Answer A. JSGT 5-22 (AC 43.13-1A)
First, locate the column on the left side of the chart representing a 28V system with a 1 volt drop. Find the horizontal line representing a wire length of 10 feet and follow it to the right until it intersects the diagonal line for 20 amps. Because the wire is in a bundle and carries a continuous current, you must be at or above curve 1 on the chart. Follow along the diagonal line representing 20 amps until it intersects curve 1. From this point, drop down vertically to the bottom of the chart. The line falls between wire sizes #12 and #14. Whenever the chart indicates a wire size between two sizes, you must select the larger wire. In this case, a #12 wire is required.

8148. Answer B. JSGT 5-22 (AC 43.13-1A)
Begin at the bottom of the chart and locate wire size #12. From this point, project a vertical line up until it intersects the diagonal line for 20 amps. Since the intersection of these two lines is above curve 2, the load requirements have been met. From the intersection, move left horizontally until you intersect the column for a 28-volt system with a 1 volt drop. The maximum length of cable that can be used is just over 25 feet. Answer (B) 26.5 feet is just over 25 feet

8149. Answer A. JSGT 5-22 (AC 65-15A)
Begin by finding the temperature along the bottom of the chart. Follow this line up to the diagonal line representing 1/8-inch 7 × 19 cable. From this point, project a line horizontally to the right side of the chart. The rigging load is 70 pounds.

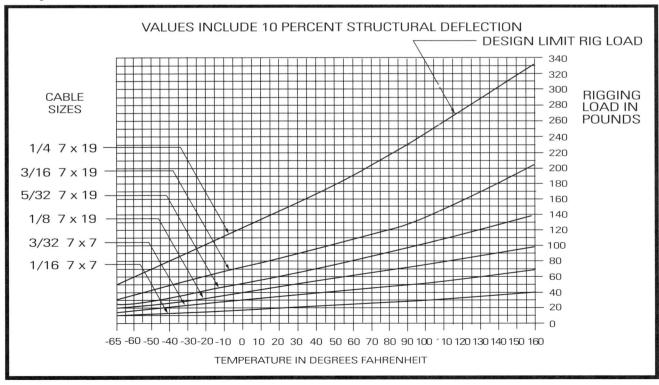

FIGURE 40.—Cable Tension Chart.

8150.   B04
(Refer to figure 40 on page 5-21) Determine the proper tension for a 3/16-inch cable (7 × 19 extra flex) if the temperature is 87°F.

A — 135 pounds.
B — 125 pounds.
C — 140 pounds.

8151.   B04
(Refer to figure 41) Determine how much fuel would be required for a 30-minute reserve operating at 2,300 RPM.

A — 25.3 pounds.
B — 35.5 pounds.
C — 49.8 pounds.

8150. Answer B. JSGT 5-22 (AC 65-9A)
Begin by locating the temperature along the bottom of the chart. From that point, move up vertically until you intersect the diagonal line representing a 3/16-inch 7 × 19 cable. From this intersection, project a line horizontally to the right side of the chart and read the rigging load. The proper rigging load is 125 pounds.

8151. Answer A. JSGT 5-24 (AC 65-9A)
To solve this problem you must first determine the specific fuel consumption. To do this, locate 2,300 RPM at the bottom of the chart and follow the vertical line up until it intersects the propeller load specific fuel consumption curve (use the full throttle curve only if full throttle operation is specified). From this point, extend a line horizontally to the right side of the chart and read the specific fuel consumption of .46 LB/BHP/HR. Now, determine the horsepower of the engine at 2,300 RPM. Again, begin at the bottom of the chart at the 2,300 RPM line and follow it up vertically to the propeller load horsepower curve. From this intersection, extend a line horizontally to the left side of the chart and read the brake horsepower (111 HP). To determine the fuel burn per hour, multiply the specific fuel consumption times the brake horsepower. The engine burns 51.06 pounds per hour (.46 × 111 = 51.06). Since this question asks for the fuel required for a 30-minute reserve, you must divide the fuel burn per hour by 2. The fuel required for a 30-minute reserve is 25.53 pounds (51.06 ÷ 2 = 25.53). Answer (A) is closest.

8152. B04

(Refer to figure 41) Determine the fuel consumption with the engine operating at cruise, 2,350 RPM.

A — 49.2 pounds per hour.
B — 51.2 pounds per hour.
C — 55.3 pounds per hour.

8152. Answer C. JSGT 5-24 (AC 65-9A)

To solve this problem, you must first determine the specific fuel consumption. To do this, locate 2,350 RPM on the bottom of the chart and follow the line up until it intersects the propeller load specific fuel consumption curve. From this intersection, extend a line to the right side of the chart and read a specific fuel consumption of .46 LB/BHP/HR. Now, determine the horsepower at 2,350 RPM. Again, begin at the bottom of the chart at the 2,350 RPM line and follow it up vertically to the propeller load horsepower curve. From this intersection, extend a line to the left side of the chart and read the brake horsepower (119 HP). To determine the fuel burn, multiply the specific fuel consumption times the brake horsepower. The engine burns 54.74 pounds per hour (.46 × 119 = 54.74). Answer (C) is closest.

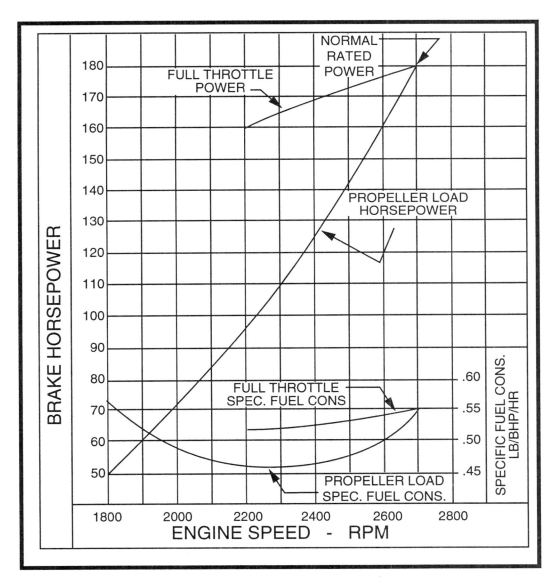

FIGURE 41.—Performance Chart.

CHAPTER 6

WEIGHT AND BALANCE

SECTION A
WEIGHING PROCEDURES

In addition to examining the procedures for weighing an aircraft, Section A of Chapter 6 discusses the theory and methodology for calculating weight and balance information. Included are terminology used in weight and balance, information on how to determine an aircraft's empty weight and center of gravity, and the importance of weight and balance for aircraft. FAA Test questions in Section A include:

8153, 8154, 8155, 8156, 8157, 8158, 8160, 8161, 8162, 8163, 8165, 8166, 8167, 8168, 8169, 8170, 8171, 8173, 8176, 8178, 8179, 8182, 8183, 8184, 8186, 8191.

8153.   C01
When computing weight and balance, an airplane is considered to be in balance when

A — the average moment arm of the loaded airplane falls within its CG range.
B — all moment arms of the plane fall within CG range.
C — the movement of the passengers will not cause the moment arms to fall outside the CG range.

8153. Answer A. JSGT 6-4 (AC 65-9A)
The CG can be thought of as the average moment arm for the aircraft. This value must fall within an allowable range to be considered safe for flight.

8154.   C01
What tasks are completed prior to weighing an aircraft to determine its empty weight?

A — Remove all items except those on the aircraft equipment list; drain fuel and hydraulic fluid.
B — Remove all items on the aircraft equipment list; drain fuel, compute oil and hydraulic fluid weight.
C — Remove all items except those on the aircraft equipment list; drain fuel and fill hydraulic reservoir.

8154. Answer C. JSGT 6-7 (AC 65-9A)
According to FAR Part 23, the empty weight of an aircraft includes all items on the aircraft's minimum equipment list, permanent ballast, unusable fuel, full hydraulics, and full oil (aircraft certified prior to March 1, 1978, are weighed with only the undrainable oil).

8155.   C01
The useful load of an aircraft consists of the

A — crew, usable fuel, passengers, and cargo.
B — crew, usable fuel, oil, and fixed equipment.
C — crew, passengers, usable fuel, oil, cargo, and fixed equipment.

8155. Answer A. JSGT 6-2 (AC 65-9A)
The useful load of an aircraft is the difference between the maximum gross weight and the empty weight. It includes items such as passengers and crew, usable fuel, and cargo.

8156.   C01
Which of the following can provide the empty weight of an aircraft if the aircraft's weight and balance records become lost, destroyed, or otherwise inaccurate?

A — Reweighing the aircraft.
B — The applicable Aircraft Specification or Type Certificate Data Sheet.
C — The applicable flight manual or pilot's operating handbook.

8156. Answer A. JSGT 6-7 (AC 65-9A)
If an aircraft's weight and balance records become lost or destroyed, the actual empty weight can only be established by reweighing the aircraft. Answer (B) is wrong because an aircraft's type certificate data sheet only lists maximum weights and not an empty weight, and answer (C) is wrong because an aircraft's flight manual contains an average empty weight for that model aircraft.

8157.   C01
In the theory of weight and balance, what is the name of the distance from the fulcrum to an object?

A — Lever arm.
B — Balance arm.
C — Fulcrum arm.

8157. Answer A. JSGT 6-5 (AC 65-9A)
In the theory of weight and balance, the name given to the distance from the fulcrum to any object is the lever arm. The lever arm is multiplied by the weight of the object to find the moment. In actual aircraft weight and balance problems, the arm is the distance from the datum to an item.

8158.   C01
(1) Private aircraft are required by regulations to be weighed periodically.
(2) Private aircraft are required to be weighed after making any alteration.
Regarding the above statements,

A — neither No. 1 nor No. 2 is true.
B — only No. 1 is true.
C — only No. 2 is true.

8158. Answer A. JSGT 6-3 (AC 65-9A)
Both of these statements are false. There is no regulation that requires private aircraft to be reweighed. Generally, after alterations, the weight changes are calculated and recorded in the aircraft's records. However, it's good operating practice to reweigh an aircraft whenever alterations may substantially affect an aircraft's weight and balance.

8160.   C01
To obtain useful weight data for purposes of determining the CG, it is necessary that an aircraft be weighed

A — in a level flight attitude.
B — with all items of useful load installed.
C — with at least minimum fuel (1/12-gallon per METO horsepower) in the fuel tanks.

8160. Answer A. JSGT 6-11 (AC 65-9A)
When an aircraft is weighed, it must be in a level flight attitude for the scale readings to be accurate. Weigh points, level points, and proper procedures for weighing are usually contained in the Type Certificate Data Sheets and are provided by the manufacturer's maintenance manual.

8161.   C01
What type of measurement is used to designate arm in weight and balance computation?

A — Distance.
B — Weight.
C — Weight/distance.

8161. Answer A. JSGT 6-5 (AC 65-9A)
Arm is the horizontal distance in inches or feet from the reference datum to an item of equipment.

8162.   C01
What determines whether the value of the moment is preceded by a plus (+) or a minus (-) sign in aircraft weight and balance?

A — The location of the weight in reference to the datum.
B — The result of a weight being added or removed and its location relative to the datum.
C — The location of the datum in reference to the aircraft CG.

8162. Answer A. JSGT 6-5 (AC 65-9A)
The location of items in an aircraft are given as positive or negative values according to their location relative to the reference datum. Items forward of the datum have a negative arm and items aft of the datum have a positive arm.

Weight and Balance 6-3

8163.   C01
The maximum weight of an aircraft is the

A — empty weight plus crew, maximum fuel, cargo, and baggage.
B — empty weight plus crew, passengers, and fixed equipment.
C — empty weight plus useful load.

8163. Answer C. JSGT 6-3 (AC 65-9A)
The maximum weight of an aircraft is the maximum authorized weight of the aircraft and its contents. This amounts to the empty weight of the aircraft plus the useful load.

8165.   C01
What should be clearly indicated on the aircraft weighing form?

A — Minimum allowable gross weight.
B — Weight of unusable fuel.
C — Weighing points.

8165. Answer C. JSGT 6-11 (AC 65-9A)
The weighing points should be clearly indicated on the aircraft weighing forms because the arm values used for the scale readings are based on these locations.

8166.   C01
If the reference datum line is placed at the nose of an airplane rather than at the firewall or some other location aft of the nose,

A — all measurement arms will be in negative numbers.
B — all measurement arms will be in positive numbers.
C — measurement arms will be in both positive and negative numbers.

8166. Answer B. JSGT 6-5 (AC 65-9A)
All measurement arms aft of the datum carry a positive (+) value and arms in front of the datum carry a negative (−) value. When the datum is located in front of the aircraft, all measurement arms are positive.

8167.   C01
Zero fuel weight is the

A — dry weight plus the weight of full crew, passengers, and cargo.
B — basic operating weight without crew, fuel, and cargo.
C — maximum permissible weight of a loaded aircraft (passengers, crew, and cargo) without fuel.

8167. Answer C. JSGT 6-2 (AC 65-9A)
Zero fuel weight is the maximum permissible weight of a loaded aircraft, including the payload (passengers, crew, and cargo), but excluding the fuel load.

8168.   C01
The empty weight of an airplane is determined by

A — adding the net weight of each weighing point and multiplying the measured distance to the datum.
B — subtracting the tare weight from the scale reading and adding the weight of each weighing point.
C — multiplying the measured distance from each weighing point to the datum times the sum of scale reading less the tare weight.

8168. Answer B. JSGT 6-13 (AC 65-9A)
Tare weight includes the weight of extra items on the weighing scale that are not part of the aircraft. To determine the empty weight of an aircraft being weighed, add the weight of each scale reading and subtract out the tare weight.

8169.   C01
When dealing with weight and balance of an aircraft, the term "maximum weight" is interpreted to mean the maximum

A — weight of the empty aircraft.
B — weight of the useful load.
C — authorized weight of the aircraft and its contents.

8169. Answer C. JSGT 6-3 (AC 65-9A)
Maximum weight is the maximum authorized weight of the aircraft and its contents. This value may be found in the Aircraft Specifications or Type Certificate Data Sheet.

8170. C02

The useful load of an aircraft is the

A — difference between the maximum gross weight and empty weight.
B — difference between the net weight and total weight.
C — sum of the empty weight and the maximum gross weight.

8171. C02

When determining the empty weight of an aircraft, certificated under current airworthiness standards (FAR Part 23), the oil contained in the supply tank is considered

A — a part of the empty weight.
B — a part of the useful load.
C — the same as the fluid contained in the water injection reservoir.

8173. C02

The maximum weight as used in weight and balance control of a given aircraft can normally be found

A — by adding the weight of full fuel, pilot, passengers, and maximum allowable baggage to the empty weight.
B — in the Aircraft Specification or Type Certificate Data Sheet.
C — by adding the empty weight and payload.

8176. C02

The amount of fuel used for computing empty weight and corresponding CG is

A — empty fuel tanks.
B — unusable fuel.
C — the amount of fuel necessary for 1/2 hour of operation.

8178. C02

As weighed, the total empty weight of an aircraft is 5,862 pounds with a moment of 885,957. However, when the aircraft was weighed, 20 pounds of alcohol were on board at +84 and 23 pounds of hydraulic fluid were in a tank located at +101. What is the empty weight CG of the aircraft?

A — 150.700.
B — 151.700.
C — 151.365.

8170. Answer A. JSGT 6-2 (AC 65-9A)
The useful load of an aircraft is determined by subtracting the empty weight from the allowable gross weight.

8171. Answer A. JSGT 6-7 (AC 65-9A)
Under current airworthiness standards (FAR Part 23), full oil is considered part of the empty weight of an aircraft. Until March 1, 1978, empty weight included only undrainable oil.

8173. Answer B. JSGT 6-7 (AC 65-9A)
The maximum weight of a particular model of aircraft is found in the Aircraft Specification or Type Certificate Data Sheet. Empty weights and useful loads vary with the equipment installed on a particular aircraft. These values must be located in the permanent records for that aircraft.

8176. Answer B. JSGT 6-7 (AC 65-9A)
The empty weight of an aircraft includes only the weight of that fuel which remains in the sumps and plumbing of the aircraft and is termed unusable fuel.

8178. Answer C. JSGT 6-6 (AC 65-9A)
The hydraulic fluid is part of the empty weight and may be ignored. Alcohol for alcohol/water injection is not part of the aircraft's empty weight and, therefore, must be subtracted out to determine the empty weight. It is typically easier to solve this type of problem if you enter the information into a table.

Item	Weight	Arm	Moment
Aircraft	5,862		885,957
Alcohol	−20	84	−1,680
Total	5,842		884,277

With the weight and moment of the alcohol subtracted out, divide the total moment by the total weight. The aircraft's empty weight CG is 151.365 inches (884,277 ÷ 5,852 = 151.365.)

8179.   C02
Two boxes which weigh 10 pounds and 5 pounds are placed in an airplane so that their distance aft from the CG are 4 feet and 2 feet respectively. How far forward of the CG should a third box, weighing 20 pounds, be placed so that the CG will not be changed?

A — 3 feet.
B — 2.5 feet.
C — 8 feet.

8182.   C02
If a 40-pound generator applies +1400 inch-pounds to a reference axis, the generator is located

A — –35 from the axis.
B — +35 from the axis.
C — +25 from the axis.

8183.   C02
In a balance computation of an aircraft from which an item located aft of the datum was removed, use

A — (–) weight × (+) arm (–) moment.
B — (–) weight × (–) arm (+) moment.
C — (+) weight × (–) arm (–) moment.

8184.   C02
Datum is forward of the main gear center point 30.24 in.
Actual distance between tail gear and
main gear center points .... 360.26 in.
Net weight at right main gear .... 9,980 lb.
Net weight at left main gear .... 9,770 lb.
Net weight at tail gear .... 1,970 lb.

These items were in the aircraft when weighed:

1. Lavatory water tank full (34 pounds at +352).
2. Hydraulic fluid (22 pounds at 8).
3. Removable ballast (146 pounds at +380).

What is the empty weight CG of the aircraft described above?

A — 62.92 inches.
B — 60.31 inches.
C — 58.54 inches.

8179. Answer B. JSGT 6-6 (AC 65-9A)
For the center of gravity to remain the same, the moments forward and aft of the CG must be equal. The two boxes added aft of the CG have moments of 40 foot/pounds (10 lbs × 4 ft = 40 ft/lbs) and 10 foot/pounds (5 lbs × 2 ft = 10 ft/lbs). This is a total moment of 50 foot/pounds. The box loaded forward must also have a moment of 50 foot/pounds to maintain balance. To determine the distance forward of the CG to maintain balance, divide 50 foot/pounds by 20 pounds. This results in a distance of 2.5 feet (50 ft/lbs ÷ 20 lbs = 2.5 ft).

8182. Answer B. JSGT 6-6 (AC 65-9A)
This question requires the application of the formula, Moment ÷ Weight = Arm. Substituting the values given, the arm is +35 inches from the axis (1,400 ÷ 40 = +35).

8183. Answer A. JSGT 6-5 (AC 65-9A)
Any time you remove something from an aircraft, the weight is subtracted (–), and any time an item is aft of the datum, it has a positive (+) arm. Therefore, a (–) weight and a (+) arm yield a (–) moment.

8184. Answer B. JSGT 6-6 (AC 65-9A)
It is typically easier to solve this type of problem if you enter the information into a table.

Item	Weight	Arm	Moment
Tail Wt.	1,970	390.50	769,285.0
L.M. Wt.	9,770	30.24	295,444.8
R.M. Wt.	9,980	30.24	301,795.2
Water	–34	352.00	–11,968.0
Ballast	–146	380.00	–55,480.0
Total	21,540		1,299,077.0

Lavatory water and removable ballast are not part of the aircraft empty weight and must be removed. Therefore, you must subtract their weights and moments to determine the empty weight CG. On the other hand, full hydraulic fluid is considered part of the aircraft empty weight and, therefore, is not used for this calculation. To calculate the CG, divide the total moment by the total weight. The CG is 60.31 inches (1,299,077.0 ÷ 21,540 = 60.31).

8186. C02
When an empty aircraft is weighed, the combined net weight at the main gears is 3,540 pounds with an arm of 195.5 inches. At the nose gear, the net weight is 2,322 pounds with an arm of 83.5 inches. The datum line is forward of the nose of the aircraft. What is the empty CG of the aircraft?

A — 151.1.
B — 155.2.
C — 146.5.

8186. Answer A. JSGT 6-6 (AC 65-9A)
It is typically easier to solve this type of problem if you enter the information into a table. The position of the datum has no bearing on this problem.

Item	Weight	Arm	Moment
Nose	2,322	83.5	193,887
Mains	3,540	195.5	692,070
Total	5,862		885,957

Calculate the CG by dividing the total moment by the total weight. The empty weight CG is 151.14 inches (885,957 ÷ 5,862 = 151.14).

8191. C02
Find the empty weight CG location for the following tricycle-gear aircraft. Each main wheel weighs 753 pounds, nosewheel weighs 22 pounds, distance between nosewheel and main wheels is 87.5 inches, nosewheel location is +9.875 inches from datum, with 1 gallon of hydraulic fluid at −21.0 inches included in the weight scale.

A — +97.375 inches.
B — +95.61 inches.
C — +96.11 inches.

8191. Answer C. JSGT 6-13 (AC 65-9A)
It is typically easier to solve this type of problem if you enter the information into a table. Remember, hydraulic fluid is included in the empty weight of the airplane, therefore should not be subtracted out.

Item	Weight	Arm	Moment
Nose	22	9.875	217.25
Mains	1,506	97.375	146,646.75
Total	1,528		146,864.00

To determine the location of the main gear, you must add the distance between the nosewheel and main gear to the location of the nosewheel. The mains are located at 97.375 (87.5 + 9.875 = 97.375 inches). Calculate the new CG by dividing the total moment by the total weight. The empty weight CG is 96.11 inches (146,864 ÷ 1,528 = 96.11).

SECTION B
SHIFTING THE CG

As items are loaded into an aircraft, or when structural modifications are made, the center of gravity (CG) moves or shifts. At times, the CG will shift beyond its acceptable range. For this reason, you must understand how shifting weight in an aircraft changes the CG location. The FAA Test questions that discuss how the CG can shift include:

    8174, 8177, 8180, 8181, 8185, 8187, 8188, 8189, 8190.

8174. C02
An aircraft with an empty weight of 2,100 pounds and an empty weight CG + 32.5 was altered as follows:
(1) two 18-pound passenger seats located +73 were removed;
(2) structural modifications were made at +77 increasing weight by 17 pounds;
(3) a seat and safety belt weighing 25 pounds were installed at +74.5; and
(4) radio equipment weighing 35 pounds was installed at +95. What is the new empty weight CG?

A — +34.01.
B — +33.68.
C — +34.65.

8174. Answer B. JSGT 6-20 (AC 65-9A)
It is typically easier to solve this type of problem if you enter the information into a table as seen below.

Item	Weight	Arm	Moment
Aircraft	2,100	32.5	68,250.0
2 Pass seats	−36	73.0	−2,628.0
Modification	17	77.0	1,309.0
Seat/belt	25	74.5	1,862.5
Radio	35	95.0	3,325.0
Total	2,141		72,118.5

First, multiply all weights by their arms to obtain the moments. Watch the negative sign on removed items. Total the weight and moment columns, then divide the new moment by the new weight. The new CG is 33.68 inches (72,118.5 ÷ 2,141 = 33.68).

8177.   C02
An aircraft as loaded weighs 4,954 pounds at a CG of +30.5 inches. The CG range is +32.0 inches to +42.1 inches. Find the minimum weight of the ballast necessary to bring the CG within the CG range. The ballast arm is +162 inches.

A — 61.98 pounds.
B — 30.58 pounds.
C — 57.16 pounds.

8177. Answer C. JSGT 6-19 (AC 65-9A)
Apply the ballast formula to the values given in the problem:

$$\text{BALLAST} = \frac{\text{(Weight of Aircraft as loaded)(Distance out of limits)}}{\text{(Distance between ballast and desired CG)}}$$

The minimum weight of the ballast necessary to bring the CG within the CG range is 57.16 pounds.

$$\frac{(4{,}954 \times 1.5)}{130} = 57.16 \text{ pounds}$$

8180.   C02
An aircraft with an empty weight of 1,800 pounds and an empty weight CG of +31.5 was altered as follows: (1) two 15-pound passenger seats located at +72 were removed; (2) structural modifications increasing the weight 14 pounds were made at +76; (3) a seat and safety belt weighing 20 pounds were installed at +73.5; and (4) radio equipment weighing 30 pounds was installed at +30.
What is the new empty weight CG?

A — +30.61.
B — +31.61.
C — +32.69.

8180. Answer B. JSGT 6-20 (AC 65-9A)
It is typically easier to solve this type of problem if you enter the information into a table.

Item	Weight	Arm	Moment
Aircraft	1,800	31.5	56,700
Seats	–30	72.0	–2,160
Mod	14	76.0	1,064
Seat/belt	20	73.5	1,470
Radio	30	30.0	900
Total	1,834		57,974

First multiply all weights by their arms to obtain the moments. Watch the negative sign on removed items. Total the weight and moment columns and divide the new moment by the new weight. The new CG is 31.61 inches (57,974 ÷ 1,834 = 31.61).

8181.   C02
An aircraft had an empty weight of 2,886 pounds with a moment of 101,673.78 before several alterations were made. The alterations included:
(1) removing two passengers seats (15 pounds each) at +71;
(2) installing a cabinet (97 pounds) at +71;
(3) installing a seat and safety belt (20 pounds) at +71; and
(4) installing radio equipment (30 pounds) at +94.
The alterations caused the new empty weight CG to move

A — 1.62 inches aft of the original empty weight CG.
B — 2.03 inches forward of the original empty weight CG.
C — 2.03 inches aft of the original empty weight CG.

8181. Answer A. JSGT 6-20 (AC 65-9A)
It is typically easier to solve this type of problem if you enter the information into a table.

Item	Weight	Arm	Moment
Aircraft	2,886	35.23	101,673.78
Seats	-30	71.00	–2,130.00
Cabinet	97	71.00	6,887.00
Seat/belt	20	71.00	1,420.00
Radio	30	94.00	2,820.00
Total	3,003		110,670.78

The old CG is calculated by dividing the original moment (101,673.78) by the original wieght (2,886 lbs). The original CG was 35.23 (101,673.78 ÷ 2,886 = 35.23). The new CG is calculated by dividing the new moment (110,670.78) by the new weight (3,003 lbs). Because the new CG (36.85) is a larger number than the original (35.23), you know that the CG has shifted to the rear (aft). Subtracting the original CG from the new CG results in a difference of 1.62 inches (36.85 – 35.23 = 1.62). Therefore, the CG shifted aft 1.62 inches.

8185.   C02
When making a rearward weight and balance check to determine that the CG will not exceed the rearward limit during extreme conditions, the items of useful load which should be computed at their minimum weights are those located forward of the

A — forward CG limit.
B — datum.
C — rearward CG limit.

8187.   C02
An aircraft with an empty weight of 1,500 pounds and an empty weight CG of +28.4 was altered as follows:
(1) two 12-pound seats located at +68.5 were removed;
(2) structural modifications weighing +28 pounds were made at +73;
(3) a seat and safety belt weighing 30 pounds were installed at +70.5; and
(4) radio equipment weighing 25 pounds was installed at +85.
What is the new empty weight CG?

A — +23.51.
B — +31.35.
C — +30.30.

8188.   C02
The following alteration was performed on an aircraft: A model B engine weighing 175 pounds was replaced by a model D engine weighing 185 pounds at a –62.00-inch station. The aircraft weight and balance records show the previous empty weight to be 998 pounds and an empty weight CG of 13.48 inches. What is the new empty weight CG?

A — 13.96 inches.
B — 14.25 inches.
C — 12.73 inches.

8189.   C02
If the empty weight CG of an airplane lies within the empty weight CG limits,

A — it is necessary to calculate CG extremes.
B — it is not necessary to calculate CG extremes.
C — minimum fuel should be used in both forward and rearward CG checks.

8185. Answer C. JSGT 6-22 (AC 65-9A)
When making a rearward extreme loading check, you must load the items of useful load behind the rearward CG limit to their maximum, and items of useful load ahead of the rearward CG limit to their minimum.

8187. Answer C. JSGT 6-20 (AC 65-9A)
It is typically easier to solve this type of problem if you enter the information into a table.

Item	Weight	Arm	Moment
Aircraft	1,500	28.4	42,600
Seats	–24	68.5	–1,644
Mod	28	73.0	2,044
Seat/belt	30	70.5	2,115
Radio	25	85.0	2,125
Total	1,559		47,240

First, multiply all weights by their arms to obtain the moments. Watch the negative sign on removed items. Total the weight and moment columns and divide the new moment by the new weight. The new empty weight CG is 30.3 (47,240 ÷ 1,559 = 30.3).

8188. Answer C. JSGT 6-20 (AC 65-9A)
It is typically easier to solve this type of problem if you enter the information into a table.

Item	Weight	Arm	Moment
Aircraft	998	13.48	13,453.04
Engine B	–175	-62.00	10,850.00
Engine D	185	-62.00	–11,470.00
Total	1,008		12,833.04

Calculate the new CG by dividing the total moment by the total weight. Upgrading to the model D engine increased the aircraft weight by 10 pounds at station –62.00. Since more weight is added ahead of the datum, the CG shifts forward to 12.73 inches (12,833.04 ÷ 1,008 = 12.73).

8189. Answer B. JSGT 6-21 (AC 65-9A)
Adverse loading checks are a deliberate attempt to load an aircraft in a manner that will create the most critical balance condition and still remain within the aircraft's design CG limits. If the empty weight CG falls within the empty weight CG range it is unnecessary to perform this check.

Weight and Balance

8190.   C02
When computing the maximum forward loaded CG of an aircraft, minimum weights, arms, and moments should be used for items of useful load that are located aft of the

A — rearward CG limit.
B — forward CG limit.
C — datum.

8190. Answer B. JSGT 6-22 (AC 65-9A)
A forward adverse-loading check is performed to determine if it is possible to load an airplane so that its CG will fall ahead of the forward CG limit. To perform this type of check, use maximum values for items of useful load that are forward of the forward limit, and minimum values for those items located aft of the forward CG limit.

SECTION C
HELICOPTER WEIGHT AND BALANCE

As an A&P technician, you must know how to compute weight and balance information on helicopters as well as airplanes. Fortunately, the theories and methodologies used for calculating weight and balance on airplanes apply to helicopters as well. FAA Test questions in Section C include:

8164, 8172, 8175.

8164.   C01
Which statement is true regarding helicopter weight and balance?

A — Regardless of internal or external loading, lateral axis CG control is ordinarily not a factor in maintaining helicopter weight and balance.
B — The moment of tail-mounted components is subject to constant change.
C — Weight and balance procedures for airplanes generally also apply to helicopters.

8164. Answer C. JSGT 6-24 (AC 65-9A)
The same theory and methodology for calculating weight and balance applies to both helicopters and airplanes. The formula Weight × Arm = Moment applies in both cases.

8172.   C02
Improper loading of a helicopter which results in exceeding either the fore or aft CG limits is hazardous due to the

A — reduction or loss of effective cyclic pitch control.
B — Coriolis effect being translated to the fuselage.
C — reduction or loss of effective collective pitch control.

8172. Answer A. JSGT 6-24 (AC 65-9A)
Helicopters generally have a smaller CG envelope than airplanes, so extra caution must be observed during loading. Trimming of a helicopter which has less than ideal balance is done with the cyclic pitch control. In extreme out-of-balance conditions full fore or aft cyclic control may be insufficient to maintain control.

8175.   C02
The CG range in single-rotor helicopters is

A — much greater than for airplanes.
B — approximately the same as the CG range for airplanes.
C — more restricted than for airplanes.

8175. Answer C. JSGT 6-24 (AC 65-9A)
Most helicopters have a restricted CG range as compared to airplanes. In some cases, this range is less than 3 inches.

CHAPTER 7

AIRCRAFT STRUCTURAL MATERIALS

SECTION A
METALS

Section A of Chapter 7 discusses the characteristics, maintenance practices, and uses of ferrous and nonferrous metals such as steel and aluminum. The following FAA Test questions are drawn from this section:

8246, 8247, 8248, 8249, 8250, 8251, 8252, 8253, 8254, 8255, 8257, 8259, 8261, 8273, 8274, 8276, 8280, 8362.

8246.   E03
Which of the following describe the effects of annealing steel and aluminum alloys?

1. decrease in internal stress.
2. softening of the metal.
3. improve corrosion resistance.

A — 1, 2.
B — 1, 3.
C — 2, 3.

8247.   E03
Which heat-treating process of metal produces a hard, wear-resistant surface over a strong, tough core?

A — Case hardening.
B — Annealing.
C — Tempering.

8248.   E03
Which heat-treating operation would be performed when the surface of the metal is changed chemically by introducing a high carbide or nitride content?

A — Tempering.
B — Normalizing.
C — Case hardening.

8249.   E03
Normalizing is a process of heat treating

A — aluminum alloys only.
B — iron-base metals only.
C — both aluminum alloys and iron-base metals.

8246. Answer A. JSGT 7-8 (AC 65-9A)
The process of annealing softens a metal and decreases its internal stresses. Answers (B) and (C) are incorrect because annealing does not improve a metal's corrosion resistance.

8247. Answer A. JSGT 7-16 (AC 65-9A)
Low-carbon and low-alloy steels may be case-hardened to give a wear resistant surface, and, at the same time, a tough internal core. This process is used on aircraft crankshafts and camshafts as well as other parts.

8248. Answer C. JSGT 7-16 (AC 65-9A)
Case-hardening is usually accomplished in one of two ways. The first is a process known as carburizing, in which controlled amounts of carbon are added to the surface of the steel to form carbides. The second method, nitriding, introduces nitrogen to the surface of the steel.

8249. Answer B. JSGT 7-15 (AC 65-9A)
Normalizing is used to relieve stresses in ferrous or iron-based metal parts. It is similar to annealing, but the particles of carbon that precipitate out are not as large as those formed during annealing. Normalizing also produces a harder and stronger material than that obtained by annealing.

8250. E03

Repeatedly applying mechanical force to most metals such as rolling, hammering, bending, or twisting commonly results in a condition known as

A — tempering or drawing.
B — stress corrosion cracking.
C — cold working, strain, or work hardening.

8251. E03

The reheating of a heat treated metal, such as with a welding torch

A — has little or no effect on a metal's heat treated characteristics.
B — can significantly alter a metal's properties in the reheated area.
C — has a cumulative enhancement effect on the original heat treatment.

8252. E03

Why is steel tempered after being hardened?

A — To increase its hardness and ductility.
B — To increase its strength and decrease its internal stresses.
C — To relieve its internal stresses and reduce its brittleness.

8253. E03

What aluminum alloy designations indicate that the metal has received no hardening or tempering treatment?

A — 3003-F.
B — 5052-H36.
C — 6061-O.

8254. E03

Which material cannot be heat treated repeatedly without harmful effects?

A — Unclad aluminum alloy in sheet form.
B — 6061-T9 stainless steel.
C — Clad aluminum alloy.

8255. E03

What is descriptive of the annealing process of steel during and after it has been annealed?

A — Rapid cooling; high strength.
B — Slow cooling; low strength.
C — Slow cooling; increased resistance to wear.

8250. Answer C. JSGT 7-4 (AC 65-9A)

Mechanically working metals at temperatures below their critical range results in strain, or work hardening of the material. Strain hardening can increase a metal part's strength and hardness. However, work hardening does reduce a material's flexibility.

8251. Answer B. JSGT 7-15 (AC 65-9A)

When a part is welded, internal stresses and strains set up in the surrounding structure that can significantly alter a metal's properties. In addition, the weld itself is a cast structure whereas the surrounding material is wrought. These two types of structures have different grain sizes and, therefore, are not very compatible. To refine the grain structure as well as relieve the internal stresses, all welded parts should be normalized after fabrication.

8252. Answer C. JSGT 7-16 (AC 65-9A)

When steel is hardened by quenching, it is usually too hard and brittle for use. Furthermore, rapid quenching causes stresses within the steel. Tempering involves reheating the material to a temperature below its critical temperature to draw out some of the hardness and relieve its internal stress.

8253. Answer A. JSGT 7-9 (AC 65-9A)

The letter F following an alloy designation indicates that the metal is in an "as fabricated" condition, and has received no hardening or tempering treatment. An "H" (answer B) on the other hand, indicates the alloy was hardened while a "O" (answer C) indicates the metal was annealed.

8254. Answer C. JSGT 7-9 (AC 65-9A)

Because clad aluminum has a corrosion-resistant coating of pure aluminum over an alloy core, repeated heat-treatments can cause diffusion of the alloying agents into the pure aluminum, resulting in reduced corrosion resistance.

8255. Answer B. JSGT 7-15 (AC 65-9A)

Annealing of steel produces a fine-grained, soft metal. Annealing is accomplished by heating the metal to just above the critical temperature and cooling it very slowly in a furnace.

8257. E04
What is generally used in the construction of aircraft engine firewalls?

A — Stainless steel.
B — Chrome-molybdenum alloy steel.
C — Magnesium-titanium alloy steel.

8257. Answer A. JSGT 7-14 (AC 65-9A)
Because of its corrosion resistant properties and high temperature performance, stainless steel is frequently used in firewalls, exhaust collectors, stacks, and manifolds.

8259. E04
Alclad is a metal consisting of

A — aluminum alloy surface layers and a pure aluminum core.
B — pure aluminum surface layers on an aluminum alloy core.
C — a homogeneous mixture of pure aluminum and aluminum alloy.

8259. Answer B. JSGT 7-6 (AC 65-9A)
The terms Alclad and Pureclad are used to designate sheets of aluminum that consist of an aluminum alloy core with a layer of pure aluminum bonded on each side.

8261. E04
The Society of Automotive Engineers (SAE) and the American Iron and Steel Institute use a numerical index system to identify the composition of various steels. In the number "4130", designating chromium molybdenum steel, the first digit indicates the

A — percentage of the basic element in the alloy.
B — percentage of carbon in the alloy in hundredths of a percent.
C — basic alloying element.

8261. Answer C. JSGT 7-12 (AC 65-9A)
The first digit of an SAE number indicates an alloy's basic alloying element. Answer (A) is wrong because the percentage of the basic element in an alloy is indicated by the second digit in the SAE number. Answer (B) is incorrect because the percent of carbon in the alloy is indicated by the third and fourth digits.

8273. E04
The core material of Alclad 2024-T4 is

A — heat-treated aluminum alloy, and the surface material is commercially pure aluminum.
B — commercially pure aluminum, and the surface material is heat-treated aluminum alloy.
C — strain-hardened aluminum alloy, and the surface material is commercially pure aluminum.

8273. Answer A. JSGT 7-6 (AC 65-9A)
The term Alclad is used to identify a sheet where a pure aluminum coating is bonded to an aluminum alloy core. The designation 2024 indicates that the core is an aluminum alloy which uses copper as its primary alloying agent. The -T4 means that the sheet has been solution heat treated, followed by natural aging at room temperature.

8274. E04
The aluminum code number 1100 identifies what type of aluminum?

A — Aluminum alloy containing 11 percent copper.
B — Aluminum alloy containing zinc.
C — 99 percent commercially pure aluminum.

8274. Answer C. JSGT 7-6 (AC 65-9A)
Commercially pure aluminum is assigned a code of 1100. It has a high degree of resistance to corrosion and is easily formed into intricate shapes.

8276. E04
In the four-digit aluminum index system number 2024, the first digit indicates

A — the major alloying element.
B — the number of different major alloying elements used in the metal.
C — the percent of alloying metal added.

8276. Answer A. JSGT 7-6 (AC 65-9A)
In the four digit aluminum index system, the first digit indicates the alloy type, or the major alloying element. The 2000 series alloys have copper as their major alloying element.

8280.   **E05**
Why is it considered good practice to normalize a part after welding?

A — To relieve internal stresses developed within the base metal.
B — To increase the hardness of the weld.
C — To remove the surface scale formed during welding.

8362.   **G02**
Parts are rinsed thoroughly in hot water after they have been heat treated in a sodium and potassium nitrate bath to

A — prevent corrosion.
B — prevent surface cracking.
C — retard discoloration.

8280. Answer A. JSGT 7-15 (AC 65-9A)
Steel that is forged, welded, or machined develops stresses within the structure that could cause failure. These stresses are typically relieved by a process known as normalizing. During the normalizing process, the part is heated above its critical temperature, then allowed to cool in still air.

8362. Answer A. JSGT 7-7 (AC 43.13-1A)
Some forms of heat treatment require immersion of parts in a heated sodium or potassium nitrate bath. Sodium and potassium nitrate are salts, and salt is a highly corrosive agent. Therefore, all parts bathed in sodium or potassium nitrate must be rinsed to prevent corrosion.

SECTION B
NONMETALLIC MATERIALS

While the greatest part of an aircraft's structure is made of metal, nonmetallic structural materials find wide use in aircraft and components. Section B of Chapter 7 covers the various types of nonmetallic materials you are likely to encounter as an aviation maintenance technician. Although the information discussed in this section is valuable, currently no FAA Test questions are based on this subject.

AIRCRAFT HARDWARE

SECTION A
AIRCRAFT RIVETS

Aircraft rivets are the single most common fastener used in aircraft fabrication. Chapter 8, Section A discusses the different types of aircraft rivets as well as several of the specialized fasteners used in aviation. Although this is an important subject, there are currently no FAA Test questions based on this material.

SECTION B
AIRCRAFT FASTENERS

Chapter 8, Section B covers the various kinds of aircraft fasteners that can be removed and reinstalled. These fasteners include bolts, nuts, washers, and screws. The FAA Test questions drawn from this section include:

8256, 8258, 8260, 8262, 8263, 8264, 8265, 8266, 8267, 8268, 8269, 8270, 8271, 8272, 8275, 8277, 8451.

8256.   E04
Unless otherwise specified, torque values for tightening aircraft nuts and bolts relate to

A — clean, dry threads.
B — clean, lightly oiled threads.
C — either dry or lightly oiled threads.

8258.   E04
Unless otherwise specified or required, aircraft bolts should be installed so that the bolthead is

A — upward, or in a forward direction.
B — downward, or in a forward direction.
C — downward, or in a rearward direction.

8260.   E04
A fiber-type, self-locking nut must never be used on an aircraft if the bolt is

A — under shear loading.
B — under tension loading.
C — subject to rotation.

8256. Answer A. JSGT 8-26 (AC 65-9A)
Unless the table of torque values for a specific fastener specifies otherwise, the torque values given are for clean, dry threads.

8258. Answer A. JSGT 8-21 (AC 65-9A)
Bolts should always be placed in the direction specified by the aircraft or engine manufacturer. However, in the absence of specific instructions, bolt heads should be installed upward or in a forward direction. This helps prevent the bolt from slipping out if the nut comes off.

8260. Answer C. JSGT 8-23 (AC 65-9A)
Self-locking nuts are used on aircraft to provide tight connections which will not shake loose under severe vibration. However, fiber-type, self-locking nuts will not remain secure at joints which are subject to rotation.

8262.   E04
(Refer to figure 42) Which of the bolthead code markings shown identifies a corrosion resistant AN standard steel bolt?

A — 1.
B — 2.
C — 3.

8262. Answer C. JSGT 8-18 (AC 65-9A)
A corrosion-resistant AN standard steel bolt is indicated by a single dash on the head. The raised cross (answer A) indicates an AN standard steel bolt while a triangle with an "X" in it (answer B) indicates a close tolerance bolt.

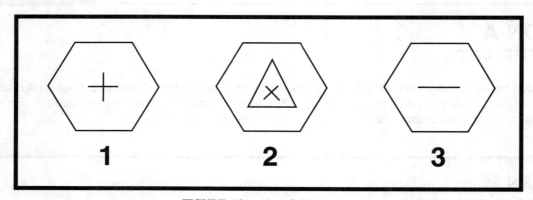

FIGURE 42.—Aircraft Hardware.

8263.   E04
Aircraft bolts with a cross or asterisk marked on the bolthead are

A — made of aluminum alloy.
B — close tolerance bolts.
C — standard steel bolts.

8263. Answer C. JSGT 8-18 (AC 65-9A)
AN standard steel bolts are identified by a cross or an asterisk on the head. An aluminum alloy bolt (answer A) is identified by two dashes on opposite sides of the bolt head and close tolerance bolts (answer B) are identified by an "X" inside a triangle.

8264.   E04
Which statement regarding aircraft bolts is correct?

A — Alloy bolts smaller than 1/4-inch diameter should not be used in primary structure.
B — When tightening castellated nuts on drilled bolts, if the cotter pin holes do not line up, it is permissible to tighten the nut up to 10 percent over recommened torque to permit alignment of the next slot with the cotter pin hole.
C — In general, bolt grip lengths should equal the material thickness.

8264. Answer C. JSGT 8-20 (AC 65-9A)
The grip length is the length of the unthreaded portion of the bolt shank. Generally the grip length should be equal to the thickness of the material being bolted together. However, bolts of slightly greater grip length may be used if washers are placed under the nut or the bolthead.

8265.   E04
Generally speaking, bolt grip lengths should be

A — equal to the thickness of the materialwhich is fastened together, plus approximatelyt one diameter.
B — equal to the thickness of the material which is fastened together.
C — one and one half times the thickness of the material which is fastened together.

8265. Answer B. JSGT 8-20 (AC 65-9A)
The grip length is the length of the unthreaded portion of the bolt shank. Generally the grip length should be equal to the thickness of the material being bolted together. However, bolts of slightly greater grip length may be used if washers are placed under the nut or the bolthead.

8266.　　E04
When the specific torque value for nuts is not given, where can the recommended torque value be found?

A — AC 43.13-2A.
B — Technical Standard Order.
C — AC 43.13-1B.

8266. Answer C. JSGT 8-26 (AC 43.13-1A)
Standard torque tables are used as a guide in tightening nuts, studs, bolts, and screws whenever specific torque values are not called out in the maintenance procedures. These tables are found in AC 43.13-1A.

8267.　　E04
(Refer to figure 43) Identify the clevis bolt illustrated.

A — 1.
B — 3.
C — 2

8267. Answer B. JSGT 8-19 (AC 65-9A)
A clevis bolt resembles a screw in that the head of the bolt is either slotted to receive a common screwdriver or recessed to receive a crosspoint screwdriver. Clevis bolts are used only in shear applications, such as connecting a control cable to a control horn.

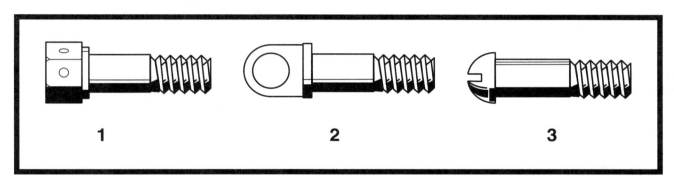

FIGURE 43.—Aircraft Hardware.

8268.　　E04
A particular component is attached to the aircraft structure by the use of an aircraft bolt and a castellated tension nut combination. If the cotter pin hole does not align within the recommended torque range, the acceptable practice is to

A — exceed the torque range.
B — tighten below the torque range
C — change washers and try again.

8268. Answer C. JSGT 8-30 (AC 43.13-1A)
When cotter pin holes don't line up with a nut castellation at the proper torque, you should vary the number of washers and try again. It is not acceptable to exceed a specified torque value (answer A).

8269.　　E04
A bolt with a single raised dash on the head is classified as an

A — AN corrosion-resistant steel bolt.
B — NAS standard aircraft bolt.
C — NAS close tolerance bolt.

8269. Answer A. JSGT 8-18 (AC 65-9A)
A single raised dash on the head of a bolt identifies an AN corrosion-resistant standard steel bolt.

8270.　　　　E04
How is a clevis bolt used with a fork-end cable terminal secured?

A — With a shear nut tightened to a snug fit, but with no strain imposed on the fork and safetied with a cotter pin.
B — With a castle nut tightened until slight binding occurs between the fork and the fitting to which it is being attached.
C — With a shear nut and cotter pin or a thin self-locking nut tightened enough to prevent rotation of the bolt in the fork.

8270. Answer A. JSGT 8-20 (AC 65-9A)
Because a clevis bolt is used only in shear loads, a shear nut is used, and is tightened only to a snug fit and secured with a cotter pin. Once installed there should be no strain imposed on the fork.

8271.　　　　E04
Where is an AN clevis bolt used in an airplane?

A — For tension and shear load conditions.
B — Where external tension loads are applied.
C — Only for shear load applications.

8271. Answer C. JSGT 8-19 (AC 65-9A)
A clevis bolt is used only where shear loads occur and never under tension loads. Clevis bolts are commonly used to attach a cable to a control horn.

8272.　　　　E04
A bolt with an × inside a triangle on the head is classified as an

A — NAS standard aircraft bolt.
B — NAS close tolerance bolt.
C — AN corrosion-resistant steel bolt.

8272. Answer B. JSGT 8-18 (AC 65-9A)
NAS close tolerance bolts are machined more accurately than a general purpose bolt. They are used in applications where a drive fit is required, and can be identified by an × inside of a triangle on the head.

8275.　　　　E04
Aircraft bolts are usually manufactured with a

A — class 1 fit for the threads.
B — class 2 fit for the threads.
C — class 3 fit for the threads.

8275. Answer C. JSGT 8-16 (AC 65-9A)
The class of a thread indicates the tolerance allowed in manufacturing. Class 1 is a loose fit, Class 2 is a free fit, Class 3 is a medium fit, and Class 4 is a close fit. Aircraft bolts are almost always manufactured to a Class 3 fit.

8277.　　　　E04
How is the locking feature of the fiber-type locknut obtained?

A — By the use of an unthreaded fiber locking insert.
B — By a fiber insert held firmly in place at the base of the load carrying section.
C — By making the threads in the fiber insert slightly smaller than those in the load carrying section.

8277. Answer A. JSGT 8-23 (AC 65-9A)
A fiber-type lock nut is a standard nut with a fiber-locking collar. This collar is not threaded and its inside diameter is smaller than the largest diameter of the threaded portion.

8451.　　　　I01
Which maintenance record entry best describes the action taken for a control cable showing approximately 20 percent wear on several of the individual outer wires at a fairlead?

A — Wear within acceptable limits, repair not necessary.
B — Removed and replaced the control cable and rerigged the system.
C — Cable repositioned, worn area moved away from fairlead.

8451. Answer A. JSGT 8-37 (AC 43.13-1A)
You should replace flexible and nonflexible cables when the individual wires in each strand appear to blend together or when the outer wires are worn 40 to 50 percent. In this example, the wear is within acceptable limits and no repair is necessary.

HAND TOOLS AND MEASURING DEVICES

SECTION A
HAND TOOLS

Section A of Chapter 9 discusses the hand tools that are typically used in the maintenance and repair of aircraft. As a technician, you must have a thorough knowledge of the the different kinds of hand tools and their correct use. There are no FAA Test questions taken from the material in this section.

SECTION B
MEASURING AND LAYOUT TOOLS

Chapter 9, Section B discusses the common measuring and layout tools used in aircraft manufacture and repair. Some of the measuring devices discussed in detail include micrometers, dial indicators, thickness gauges, telescoping gauges, and calipers. The following FAA Test questions are based on the material presented in this section:

8289, 8290, 8291, 8292, 8293, 8294, 8295, 8296, 8297, 8298, 8299, 8300, 8301, 8302, 8303, 8304, 8305, 8306, 8307.

8289.   E06
Which tool can be used to measure the alignment of a rotor shaft or the plane of rotation of a disk?

A — Dial indicator.
B — Shaft gauge.
C — Protractor.

8289. Answer A. JSGT 9-40 (AC 43.13-1A)
Dial indicators are precision measuring instruments that are used to measure end-play, and to check shaft alignments.

8290. E06
(Refer to figure 46) The measurement reading on the illustrated micrometer is

A — 0.2851.
B — 0.2911.
C — 0.2901.

8290. Answer A. JSGT 9-39 (AC 65-9A)
Each line on the barrel represents .025 inches. There are 11 lines showing, so you know the measurement is at least .275 (.025 × 11 = .275). Now look at the thimble. Each line represents .0010 inches. The line on the barrel is between the 10 and 11, which indicates the measurement is at least .010 inches longer, or .285 (.275 + .01 = .285). The number on the top of the barrel represents .0001 (ten-thousandth) inches. To identify the number of ten-thousandths, look for the line on the thimble that is aligned with a horizontal line on the barrel. In this case the line indicating 1 on the barrel is aligned with the 15 on the thimble. This means that .001 must be added to the measurement. The total measurement is .2851 (.285 + .0001 = .2851).

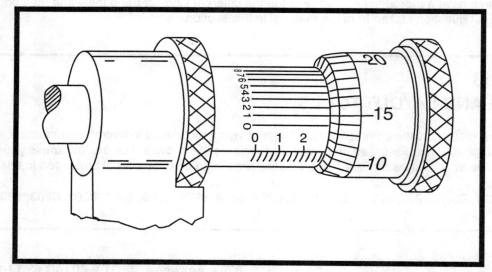

FIGURE 46.—Precision Measurement.

8291. E06
Identify the correct statement.

A — An outside micrometer is limited to measuring diameters.
B — Tools used on certificated aircraft must be an approved type.
C — Dividers do not provide a reading when used as a measuring device.

8291. Answer C. JSGT 9-34 (AC 65-9A)
Dividers are a layout tool consisting of two legs with sharp points. They are used to scribe circles and arcs, and for transferring measurements. However, dividers have no way of indicating a measurement.

Hand Tools and Measuring Devices

8292.   E06
(Refer to figure 47) What is the measurement reading on the venier caliper scale?

A — 1.411 inches.
B — 1.436 inches.
C — 1.700 inches.

8293.   E06
Which tool is used to measure the clearance between a surface plate and a relatively narrow surface being checked for flatness?

A — Depth gauge.
B — Thickness gauge.
C — Dial indicator.

8294.   E06
Which number represents the vernier scale graduation of a micrometer?

A — .00001.
B — .001.
C — .0001.

8292. Answer B. JSGT 9-39 (AC 65-9A)
On a venier caliper inches, tenths of inches, and divisions of 25-thousands of an inch are taken from the top scale. The bottom scale represents thousandths of an inch. In this example, the zero mark on the bottom scale is between the 1-inch and 2-inch mark on the top scale indicating the measurement is greater than 1 inch. The zero mark is also beyond the 4 indicating the measurement is greater than 1.4 inches (1 + .4 = 1.4). The top scale also indicates that more than 25-thousandths must be added. This results in a measurement that is at least 1.425 inches (1.4 + .025 = 1.425). To determine how many thousandths greater than 25 thousandths, identify the line on the top scale that is aligned with a line on the bottom scale and read the number of thousandths off the bottom scale. An additional 11-thousandths must be added. The total measurement of 1.436 (1.425 + .011 = 1.436) is correct.

8293. Answer B. JSGT 9-34 (AC 65-12A)
A thickness gauge, often referred to as a feeler gauge, is often used to determine the clearance between a surface plate and a surface being checked for flatness. A depth gauge (answer A) indicates the depth of something and a dial indicator (answer C) allows you to check for shaft alignment and end play.

8294. Answer C. JSGT 9-37 (AC 65-9A)
A micrometer measures with a vernier scale on the barrel to the nearest ten-thousandths of an inch or .0001.

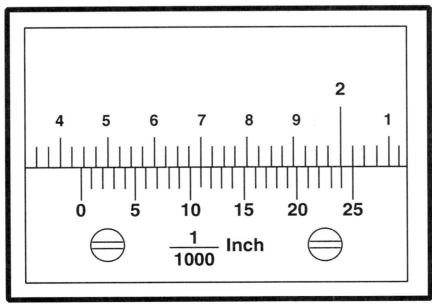

FIGURE 47.—Precision Measurement.

8295.   E06
Which tool is used to find the center of a shaft or other cylindrical work?

A — Combination set.
B — Dial indicator.
C — Micrometer caliper.

8296.   E06
(Refer to figure 48) What does the micrometer read?

A — .2974.
B — .3004.
C — .3108.

8295. Answer A. JSGT 9B (AC 65-9A)
A combination set consists of a steel scale, a stock head, a protractor head, and a centering head. The centering head is used to find the center of cylindrical pieces.

8296. Answer B. JSGT 9B (AC 65-9A)
The micrometer is read as follows:

Barrel	—	.300
Thimble	—	.000
Vernier	—	.0004
Total	—	.3004

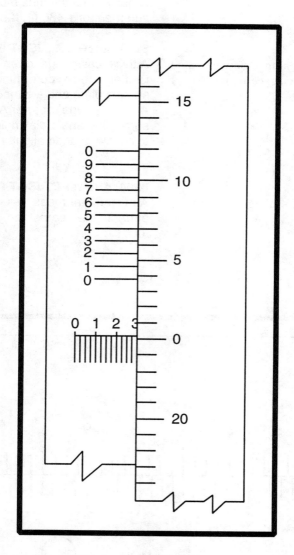

FIGURE 48.—Precision Measurement.

Hand Tools and Measuring Devices

8297. E06
If it is necessary to accurately measure the diameter of a hole approximately 1/4 inch in diameter, the mechanic should use a

A — telescoping gauge and determine the size of the hole by taking a micrometer reading of the adjustable end of the telescoping gauge.
B — 0- to 1-inch inside micrometer and read the measurement directly from the micrometer.
C — small-hole gauge and determine the size of the hole by taking a micrometer reading of the ball end of the gauge.

8297. Answer C. JSGT 9-40
Small hole gauges are used to transfer the diameter measurement of a hole to a micrometer to determine the size. Telescoping gauges (answer A) serve the same purpose, but are not generally available in sizes as small as 1/4".

8298. E06
(Refer to figure 49) The measurement reading on the micrometer is

A — .2758.
B — .2702.
C — .2792.

8298. Answer C. JSGT 9-38 (AC 65-9A)
The micrometer reading is interpreted as follows:

Barrel   — .275
Thimble — .004
Vernier — .0002
Total   — .2792

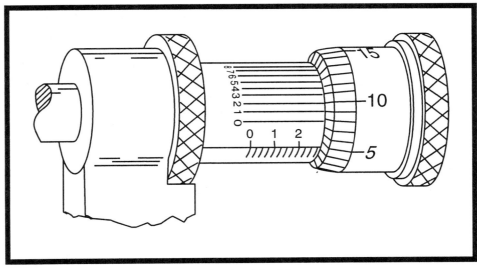

FIGURE 49.—Precision Measurement.

8299. E06
What tool is generally used to set a divider to an exact dimension?

A — Machinist scale.
B — Surface gauge.
C — Dial indicator.

8299. Answer A. JSGT 9-34 (AC 65-9A)
Dividers are a layout tool, not a precision measuring device. They are generally set using a machinist scale.

8300. E06
What tool is generally used to calibrate a micrometer or check its accuracy?

A — Gauge block.
B — Dial indicator.
C — Machinist scale.

8300. Answer A. JSGT 9-37
Special gauge blocks are generally included when micrometers are purchased. They may be used to test the micrometer for accuracy, and to calibrate when required.

**8301.	E06**
What precision measuring tool is used for measuring crankpin and main bearing journals for out-of-round wear?

A — Dial gauge.
B — Micrometer caliper.
C — Depth gauge.

**8302.	E06**
The side clearances of piston rings are measured with a

A — micrometer caliper gauge.
B — thickness gauge.
C — dial gauge.

**8303.	E06**
How can the dimensional inspection of a bearing in a rocker arm be accomplished?

A — Depth gauge and micrometer.
B — Thickness gauge and push-fit arbor.
C — Telescopic gauge and micrometer.

**8304.	E06**
The twist of a connecting rod is checked by installing push-fit arbors in both ends, supported by parallel steel bars on a surface plate. Measurements are taken between the arbor and the parallel bar with a

A — dial gauge.
B — height gauge.
C — thickness gauge.

**8305.	E06**
The clearance between the piston rings and the ring lands is measured with a

A — micrometer caliper.
B — thickness gauge.
C — depth gauge.

**8306.	E06**
What may be used to check the stem on a poppet-type valve for stretch?

A — Dial indicator.
B — Micrometer.
C — Telescoping gauge.

**8307.	E06**
Which tool can be used to determine piston pin out-of-round wear?

A — Telescopic gauge.
B — Micrometer caliper.
C — Dial indicator.

8301. Answer B. JSGT 9-34 (AC 43.13-1A)
In many cases, an outside micrometer is used to determine an out-of-round condition. To do this, take a measurement at one point and then rotate the micrometer 90 degrees and take a second measurement.

8302. Answer B. JSGT 9-34 (AC 65-12A)
Side clearance of piston rings is checked with a thickness gauge, or feeler gauge.

8303. Answer C. JSGT 9-40 (AC 65-12A)
A telescoping gauge is used to transfer the diameter of the bearing to an outside micrometer.

8304. Answer C. JSGT 9-34 (AC 65-12A)
To check for twisting in a connecting rod, install arbors in both ends of the connecting rod and lay it across parallel blocks on a surface plate. Then, check the clearance at the points where the arbors rest on the blocks with a thickness gauge.

8305. Answer B. JSGT 9-34 (AC 65-12A)
The clearance between the piston rings and the ring lands is called side clearance and is easily checked using a thickness gauge.

8306. Answer B. JSGT 9-35 (AC 65-12A)
Valve stretch is indicated by a decrease in diameter near the neck of the valve stem. This is easily measured using a micrometer caliper.

8307. Answer B. JSGT 9-35
By taking measurements with an outside micrometer at 90 degrees to each other, an out-of-round condition may be determined.

CHAPTER 10

FLUID LINES AND FITTINGS

SECTION A
RIGID FLUID LINES

Among other functions, rigid fluid lines and fittings supply fuel, hydraulic power, and breathing oxygen in modern aircraft. Therefore, as an aviation maintenance technician, it is important that you understand how to properly maintain and repair rigid tubing. Chapter 10, Section A presents important material on the rigid fluid lines you will most likely encounter. FAA Test questions based on this section include:

8192, 8193, 8194, 8195, 8196, 8198, 8199, 8200, 8203, 8204, 8205, 8206, 8207, 8208, 8210, 8212, 8213, 8214, 8215, 8217, 8452.

8192. D01
Which coupling nut should be selected for use with 1/2-inch aluminum oil lines which are to be assembled using flared tube ends and standard AN nuts, sleeves, and fittings?

A — AN-818-5.
B — AN-818-16.
C — AN-818-8.

8192. Answer C. JSGT 10-3 (AC 65-9A)
Since the coupling nut used must fit a 1/2-inch line, the final dash number in the AN-818 series coupling nut indicates the size tube that it will fit. The numbers are in 1/16 inch increments, so a -8 is equivalent to 8/16 or 1/2-inch (answer B).

8193. D01
Metal tubing fluid lines are sized by wall thickness and

A — outside diameter in 1/16 inch increments.
B — inside diameter in 1/16 inch increments.
C — outside diameter in 1/32 inch increments.

8193. Answer A. JSGT 10-13 (AC 65-9A)
All rigid metal fluid lines are sized by wall thickness and outside diameter in 1/16 inch increments. Answer (B) is incorrect because flexible fluid lines, not rigid fluid lines are sized by inside diameter. Answer (C) is wrong because no fluid lines are sized in 1/32 inch increments.

8194. D01
From the following sequences of steps, indicate the proper order you would use to make a single flare on a piece of tubing:

1. Place the tube in the proper size hole in the flaring block.
2. Project the end of the tube slightly from the top of the flaring tool, about the thickness of a dime.
3. Slip the fitting nut and sleeve on the tube.
4. Strike the plunger several light blows with a lightweight hammer or mallet and turn the plunger one-half turn after each blow.
5. Tighten the clamp bar securely to prevent slippage.
6. Center the plunger or flaring pin over the tube.

A — 1, 3, 5, 2, 4, 6.
B — 3, 1, 6, 2, 5, 4.
C — 3, 2, 6, 5, 1, 4.

8194. Answer B. JSGT 10-6 (AC 65-9A)
This question is geared to the impact-type single flaring tool. To prepare a tube for flaring, slip the fitting nut and sleeve on the tube (3) and place the tube in the proper size hole in the flaring tool (1). Center the plunger or flaring pin over the end of the tube (6). Then project the end of the tubing slightly from the top of the flaring tool, about the thickness of a dime (2), and tighten the clamp bar securely to prevent slippage (5). Make the flare by striking the plunger several light blows with a lightweight hammer or mallet (4). Turn the plunger a half turn after each blow and be sure it seats properly before removing the tube from the flaring tool.

8195. D01
Hydraulic tubing, which is damaged in a localized area to such an extent that repair is necessary, may be repaired

A — by cutting out the damaged area and utilizing a swaged tube fitting to join the tube ends.
B — only by replacing the entire tubing using the same size and material as the original.
C — by cutting out the damaged section and soldering in a replacement section of tubing.

8195. Answer A. JSGT 10-13 (AC 43.13-1A)
A severely damaged line is typically replaced; however, if the damage is localized, the line may be repaired by cutting out the damaged area and inserting a tube of the same size and material. To do this, remove the damaged area and fabricate a replacement section. The replacement section can be secured using standard unions, sleeves, and tube nuts.

8196. D01
What is an advantage of a double flare on aluminum tubing?

A — Ease of construction.
B — It is less resistant to the shearing effect of torque.
C — It is more resistant to the shearing effect of torque.

8196. Answer C. JSGT 10-6 (AC 65-9A)
A double flare should be used on soft aluminum tubing from 1/8" to 3/8" in diameter. The double flare is smoother and more concentric than a single flare, and is also more resistant to the shearing effects of torque.

8199. D01
What is the color of an AN steel flared-tube fitting?

A — Black.
B — Blue.
C — Red.

8199. Answer A. JSGT 10-8 (AC 65-9A)
For identification purposes, all AN steel fittings are colored black, and all AN aluminum alloy fittings are colored blue.

8200. D01
Select the correct statement in reference to flare fittings.

A — AC fittings are generally replacing the older AN fittings.
B — AC and AN fittings are identical except for the material composition and identifying color.
C — AN fittings can easily be identified by the shoulder between the end of the threads and the flare cone.

8200. Answer C. JSGT 10-7 (AC 65-9A)
AN fittings have replaced the AC fittings, but occasionally AC fittings are found on older aircraft. The fastest way to identify the AN fitting is by the shoulder between the end of the threads and the flare cone. The threads on the AC fitting extend to the flare cone while the AN fitting has a shoulder between the flare cone and the threads. Another way to identify AC fittings is by their gray or yellow color. Aluminum alloy AN fittings are dyed blue.

Fluid Lines and Fittings

8203.   D01
Excessive stress on fluid or pneumatic metal tubing caused by expansion and contraction due to temperature changes can best be avoided by

A — using short, straight sections of tubing between fixed parts of the aircraft.
B — not exposing the aircraft to temperature extremes or sudden changes in temperature.
C — providing suitable bends in the tubing.

8203. Answer C. JSGT 10-11 (AC 43.13-1A)
Providing suitable bends in rigid tubing will allow for expansion and contraction caused by temperature changes without creating undue stress. Answer (A) is incorrect because, if no bends are present, excessive stress can be imposed on the tubing. On the other hand, answer (B) is incorrect because it is nearly impossible to keep an aircraft from being exposed to sudden temperature changes.

8204.   D01
The material specifications for a certain aircraft require that a replacement oil line be fabricated from 3/4-inch 0.072 5052-0 aluminum alloy tubing. What is the inside dimension of this tubing?

A — 0.606 inch.
B — 0.688 inch.
C — 0.750 inch.

8204. Answer A. JSGT 10-3 (AC 65-9A)
The 3/4-inch identifies the outside diameter. The 0.072 identification is the wall thickness. To find the inside diameter you must subtract two wall thicknesses from the outside diameter. The inside diameter is 0.606 inches [0.750 − (2 × .072) = 0.606].

8205.   D01
In most aircraft hydraulic systems, two-piece tube connectors consisting of a sleeve and a nut are used when a tubing flare is required. The use of this type connector eliminates

A — the flaring operation prior to assembly.
B — the possibility of reducing the flare thickness by wiping or ironing during the tightening process.
C — wrench damage to the tubing during the tightening process.

8205. Answer B. JSGT 10-5 (AC 65-9A)
Single-piece connectors are not typically used on flared fittings because the nut tends to wipe or iron the flare as it is tightened. However, when two-piece connectors consisting of a sleeve and nut are used the nut bears on the sleeve, not the flare.

8206.   D01
Which statement about Military Standard (MS) flareless fittings is correct?

A — During installation, MS flareless fittings are normally tightened by turning the nut a specified amount after the sleeve and fitting sealing surface have made contact, rather than being torqued.
B — MS flareless fittings should not be lubricated prior to assembly.
C — MS flareless fittings must be tightened to a specific torque.

8206. Answer A. JSGT 10-9 (AC 65-9A)
Prior to installation of a new flareless tube assembly, the fitting must be preset. To do this, flareless fittings are tightened by turning the nut a specified amount after contact is made between the sleeve and fitting surface (answer A). For aluminum alloy tubing up to and including 1/2″, tighten the nut from 1 to 1-1/6 turns. For steel tubing and aluminum alloy tubing over 1/2″, tighten from 1-1/6 to 1-1/2 turns. When re-attaching a flareless fitting that has been preset, you should tighten the nut by hand until it begins to bottom out. Once this occurs, tighten the nut with a wrench an additional one-sixth turn. If the connections leaks, it is permissible to tighten the nut an additional one-sixth turn.

8207.   D01
When flaring aluminum tubing for use with AN fittings, the flare angle should be

A — 37°.
B — 35°.
C — 45°.

8207. Answer A. JSGT 10-5 (AC 65-9A)
The flare angle used on aircraft fittings is 37°. Under no circumstances should you use an automotive type flaring tool since they produce a 45° flare angle. Use caution when selecting flaring tools as this difference is difficult to detect by eye.

8208.   D01
Scratches or nicks on the straight portion of aluminum alloy tubing may be repaired if they are no deeper than

A — 20 percent of the wall thickness.
B — 1/32 inch or 20 percent of wall thickness, whichever is less.
C — 10 percent of the wall thickness.

8210.   D01
A scratch or nick in aluminum alloy tubing can be repaired by burnishing, provided the scratch or nick does not

A — appear in the heel of a bend in the tube.
B — exceed 20 percent of the wall thickness of the tube.
C — exceed 10 percent of the tube diameter.

8212.   D01
Which tubings have the characteristics (high strength, abrasion resistance) necessary for use in a high-pressure (3,000 PSI) hydraulic system for operation of landing gear and flaps?

A — 2024-T or 5052-0 aluminum alloy.
B — Corrosion-resistant steel annealed or 1/4H.
C — 1100-1/2H or 3003-1/2H aluminum alloy.

8213.   D01
When installing bonded clamps to support metal tubing,

A — paint removal from tube is not recommended as it will inhibit corrosion.
B — paint clamp and tube after clamp installation to prevent corrosion.
C — remove paint or anodizing from tube at clamp location.

8214.   D01
In a metal tubing installation,

A — rigid straight line runs are preferable.
B — tension is undesirable because pressurization will cause it to expand and shift.
C — a tube may be pulled in line if the nut will start on the threaded coupling.

8215.   D01
A gas or fluid line marked with the letters PHDAN

A — is a high-pressure line. The letters mean Pressure High, Discharge at Nacelle.
B — is carrying a substance which may be dangerous to personnel.
C — must be made of a nonphosphorous metal.

8208. Answer C. JSGT 10-12 (AC 65-9A)
Scratches or nicks in aluminum alloy tubing that are no deeper than 10 percent of the wall thickness and not in the heel of a bend may be repaired by burnishing with hand tools.

8210. Answer A. JSGT 10-12 (AC 65-9A)
Scratches or nicks in aluminum tubing that are no deeper than 10 percent of the wall thickness and not in the heel of a bend may be repaired by burnishing with hand tools.

8212. Answer B. JSGT 10-2 (AC 65-9A)
Corrosion resistant steel (stainless), either annealed or 1/4-hard, is used extensively in high-pressure hydraulic systems for the operation of landing gear, flaps, brakes, and the like.

8213. Answer C. JSGT 10-12 (AC 65-9A)
Bonded clamps are used to secure metal hydraulic, fuel, and oil lines in place. Before installing these clamps you should remove any paint or anodizing from that portion of the tube the bonding clamp makes contact with. Any coatings on the line could defeat the purpose of the bonding clamp.

8214. Answer B. JSGT 10-10 (AC 65-9A)
Metal tubing should never be installed under tension. All rigid tubing should have at least one bend between the fittings to absorb vibrations, strains from vibrations, and strain from dimensional changes in the aircraft structure.

8215. Answer B. JSGT 10-13 (AC 65-9A)
Lines containing physically dangerous materials, such as oxygen, nitrogen, or Freon may be marked PHDAN (PHysical DANger).

Fluid Lines and Fittings 10-5

8217.   D01
(1) Bonded clamps are used for support when installing metal tubing.
(2) Unbonded clamps are used for support when installing wiring.
Regarding the above statements,

A — only No. 1 is true.
B — both No. 1 and No. 2 are true.
C — neither No. 1 nor No. 2 are true.

8217. Answer B. JSGT 10-12 (AC 65-9A)
Both statements are correct. Bonded clamps are used to secure metal hydraulic, fuel, and oil lines. Unbonded clamps are typically used to secure wiring.

8452.   I01
Which maintenance record entry best describes the action taken for a .125-inch deep dent in a straight section of 1/2-inch aluminum alloy tubing?

A — Dented section removed and replaced with identical new tubing flared to 45°.
B — Dent within acceptable limits, repair not necessary.
C — Dented section removed and replaced with identical new tubing flared to 37°.

8452. Answer C. JSGT 10-12 (AC 43.13-1A)
Dents less than 20 percent of the diameter of the tube are acceptable in straight sections. However, .125 is 25 percent of 1/2", so this dent must be repaired.

SECTION B
FLEXIBLE FLUID LINES

Section B of Chapter 10 contains information on flexible fluid lines. Because many components in aircraft move or vibrate, flexible fluid lines are essential components in the construction of aircraft. The following FAA Test questions are drawn from this section:

8197, 8201, 8202, 8209, 8211, 8218.

8197.   D01
A certain amount of slack must be left in a flexible hose during installation because, when under pressure, it

A — expands in length and diameter.
B — expands in length and contracts in diameter.
C — contracts in length and expands in diameter.

8197. Answer C. JSGT 10-17 (AC 65-9A)
When under pressure, flexible hose contracts in length and expands in diameter. Because of this, flexible hose installations should have 5 to 8 percent of the total length left as slack to allow for freedom of movement.

8198.   D01
The term "cold flow" is generally associated with

A — vaporizing fuel.
B — rubber hose.
C — welding and sheet metal.

8198. Answer B. JSGT 10-10 (AC 65-9A)
The term cold flow describes the deep, permanent impressions in rubber hose produced by the pressure of hose clamps or supports.

8201.   D01
Flexible lines must be installed with

A — enough slack to allow maximum flexing during operation
B — a slack of at least 10 to 12 percent of the length.
C — a slack of 5 to 8 percent of the length.

8201. Answer C. JSGT 10-17 (AC 65-9A)
Flexible fluid lines must have between 5 and 8 percent slack to allow for changes in dimension caused by pressure, vibration, and expansion and contraction of the airframe.

8202. **D01**
The maximum distance between end fittings to which a straight hose assembly is to be connected is 50 inches. The minimum hose length to make such a connection should be

A — 54-1/2 inches.
B — 51 inches.
C — 52-1/2 inches.

8202. Answer C. JSGT 10-17 (AC 65-9A)
Flexible fluid lines must have between 5 and 8 percent slack to allow for changes in dimension caused by pressure, vibration, and expansion and contraction of the airframe. Therefore, you must add 2-1/2 inches to the length of the hose (50 ft. × .05 = 2.5 ft.). The minimum hose length is 52-1/2 inches.

8209. **D01**
Flexible hose used in aircraft systems is classified in size according to the

A — outside diameter.
B — wall thickness
C — inside diameter.

8209. Answer C. JSGT 10-16 (AC 65-9A)
The size of flexible hose is determined by its inside diameter, whereas rigid fluid lines are determined by outside diameter.

8211. **D01**
Which of the following hose materials are compatible with phosphate-ester base hydraulic fluids?

1. Butyl.
2. Teflon.
3. Buna-N.
4. Neoprene.

A — 1 and 2.
B — 2 and 4.
C — 1 and 3.

8211. Answer B. JSGT 10-14 (AC 65-9A)
A teflon hose can be used to carry fuels, lubricating oils, coolants, or solvents.

8218. **D01**
A 3/8 inch aircraft high pressure flexible hose as compared to 3/8 inch metal tubing used in the same system will

A — have about the same OD.
B — have equivalent flow characteristics.
C — usually have interchangeable applications.

8218. Answer B. JSGT 10-18 (AC 65-9A)
Flexible hose is sized by inside diameter while rigid tubing is sized by outside diameter. Based on this, the size of the opening that fluid flows through is nearly the same and produces equivalent flow characteristics for identically sized rigid and flexible tubing. Answer (A) is incorrect because, in order to provide the same strength requirements, the wall thickness of flexible tubing must be greater than the wall thickness of rigid tubing and, therefore, the outside diameter of the two will be different. Answer (C) is wrong because the applications in which the flexible hose and rigid tubing are used are not interchangeable.

CHAPTER 11

NONDESTRUCTIVE TESTING

SECTION A
VISUAL INSPECTIONS

Visual inspections are the most common and least expensive method of inspecting aircraft and their components. Chapter 11, Section A discusses the various types of visual inspections including dye penetrant, and magnetic particle inspections. The FAA Test questions covering this information include:

8219, 8220, 8223, 8225, 8226, 8228, 8229, 8230, 8231, 8232, 8233, 8234, 8235, 8236, 8237, 8238, 8239, 8240, 8241, 8242, 8243, 8244, 8278, 8279, 8281, 8282, 8283, 8284, 8285, 8286, 8287, 8288.

8219.　　　E01
Magnetic particle inspection is used primarily to detect

A — distortion.
B — deep subsurface flaws.
C — flaws on or near the surface.

8219. Answer C. JSGT 11-7 (AC 65-9A)
Because the magnetic fields generated within a material are weak, magnetic particle inspection is primarily used to detect surface flaws in ferromagnetic materials such as iron and steel. The magnetic field produced by distortions (answer A) and deep subsurface flaws (answer B) are typically not strong enough to attract the indicating medium.

8220.　　　E01
Liquid penetrant inspection methods may be used on which of the following?

1. porous plastics.
2. ferrous metals.
3. nonferrous metals.
4. smooth unpainted wood.
5. nonporous plastics.

A — 2, 3, 4.
B — 1, 2, 3.
C — 2, 3, 5.

8220. Answer C. JSGT 11-4 (AC 65-9A)
Penetrant inspection is an effective nondestructive testing method for detecting defects that are open to the surface in materials such as ferrous metals, nonferrous metals, and nonporous materials. Even if the defects are too small for visual or other methods of detection, as long as the defect is open to the surface, dye penetrant inspection is effective. The magnetic characteristics of the material being checked (answers A and B) have no bearing as to the effectiveness of dye penetrant inspection.

8223.　　　E01
What method of magnetic particle inspection is used most often to inspect aircraft parts for invisible cracks and other defects?

A — Residual.
B — Inductance.
C — Continuous.

8223. Answer C. JSGT 11-10 (AC 65-9A)
In the continuous method of magnetic particle inspection, a part is magnetized and the indicating medium applied while the magnetizing force is maintained. The continuous method is used in practically all circular and longitudinal magnetization procedures because it provides greater sensitivity than the residual method (answer A), particularly in locating subsurface defects.

8225.   E01
The testing medium that is generally used in magnetic particle inspection utilizes a ferromagnetic material that has

A — high permeability and low retentivity.
B — low permeability and high retentivity.
C — high permeability and high retentivity.

8225. Answer A. JSGT 11-9 (AC 65-9A)
In addition to being ferromagnetic, and finely divided, the testing medium used in magnetic particle inspection must conduct lines of magnetic flux easily (high permeability), and not retain a high degree of magnetism after the magnetizing current is removed (low retentivity).

8226.   E01
Which statement relating to the residual magnetizing inspection method is true?

A — Subsurface discontinuities are made readily apparent.
B — It is used in practically all circular and longitudinal magnetizing procedures.
C — It may be used only with steels which have been heat treated for stressed applications.

8226. Answer C. JSGT 11-10 (AC 65-9A)
The residual inspection procedure is used only with steels which have been heat treated for stressed applications. The continuous magnetizing inspection method is used on almost all circular and longitudinal magnetizing procedures (answer B). In either case, subsurface discontinuities are not made apparent using either method (answer A).

8228.   E02
What two types of indicating mediums are available for magnetic particle inspection?

A — Wet and dry process materials.
B — High retentivity and low permeability material.
C — Iron and ferric oxides.

8228. Answer A. JSGT 11-9 (AC 65-9A)
Magnetic particle inspection media are either wet or dry process materials. The dry process uses iron oxide powder poured or sprayed on the part being tested. The wet process requires mixing the oxide powder with kerosene or some other light oil. The part being inspected is then immersed in the oxide-oil bath. Answer (B) is incorrect because the media used in magnetic particle inspection must have a low retentivity and a high permeability.

8229.   E02
Which of the folling materials may be inspected using the magnetic particle inspection method?

1. Magnesium alloys.   A — 1,2,3
2. Aluminum alloys.   B — 1,2,4,5
3. Iron alloys.   C — 3
4. Copper alloys.
5. Zinc alloys.

8229. Answer C. JSGT 11-7 (AC 65-9A)
Magnetic particle inspection is limited to use on ferromagnetic materials such as iron and steel. Alloys containing primarily magnesium (answer A) or aluminum (answer B) are non-magnetic and, therefore, cannot be inspected using magnetic particle inspection.

8230.   E02
One way a part may be demagnetized after magnetic particle inspection is by

A — subjecting the part to high voltage, low amperage ac.
B — slowly moving the part out of an ac magnetic field of sufficient strength.
C — slowly moving the part into an ac magnetic field of sufficient strength.

8230. Answer B. JSGT 11-12 (AC 65-9A)
If a part is subjected to an AC magnetizing force and then slowly removed from the force while the current is still flowing, the domains within the material become disorganized and magnetization decreases.

8231.   E02
Which type crack can be detected by magnetic particle inspection using either circular or longitudinal magnetization?

A — 45°.
B — Longitudinal.
C — Transverse.

8231. Answer A. JSGT 11-8 (AC 65-9A)
In order to locate a defect using magnetic particle inspection, it is essential that the lines of flux pass approximately perpendicular to the defect. Circular magnetization locates defects running approximately parallel to the axis of the part, while longitudinal magnetization locates defects running approximately 90° to the axis. If a defect is 45° to the axis of the part, either method should detect it.

8232. E02
Which of the following methods may be suitable to use to detect cracks open to the surface in aluminum forgings and castings?

1. Dye penetrant inspection.
2. Magnetic particle inspection.
3. Metallic ring (coin tap) inspection.
4. Eddy current inspection.
5. Ultrasonic inspection.
6. Visual inspection.

A — 1, 4, 5, 6.
B — 1, 2, 4, 5, 6.
C — 1, 2, 3, 4, 5, 6.

8232. Answer A. JSGT 11-4 (AC 65-9A)
If a crack is open to the surface on an aluminum forging, the crack can be detected through the use of dye penetrant, eddy current, ultrasonic, or visual inspections. Answers (B) and (C) are incorrect because magnetic particle inspection can only be used on ferrous metals and aluminum is not a ferrous metal. In addition, the metallic ring, or coin tap inspection is used primarily on composite structures.

8233. E02
To detect a minute crack using dye penetrant inspection usually requires

A — that the developer be applied to a flat surface.
B — a longer-than-normal penetrating time.
C — the surface to be highly polished.

8233. Answer B. JSGT 11-6 (AC 65-9A)
Dye penetrant inspection is based on capillary attraction. In other words, dye is drawn into a crack or defect until developer is applied making the defect visible. A good rule to remember when using dye penetrant is that small defects require a longer penetrating time.

8234. E02
When checking an item with the magnetic particle inspection method, circular and longitudinal magnetization should be used to

A — reveal all possible defects.
B — evenly magnetize the entire part.
C — ensure uniform current flow.

8234. Answer A. JSGT 11-8 (AC 65-9A)
Circular magnetization reveals defects running approximately parallel to the axis of the part, while longitudinal magnetization reveals defects that are approximately 90° to the axis. For this reason, it is best to use both methods to reveal all possible defects.

8235. E02
In magnetic particle inspection, a flaw that is perpendicular to the magnetic field flux lines generally causes

A — a large disruption in the magnetic field.
B — a minimal disruption in the magnetic field.
C — no disruption in the magnetic field.

8235. Answer A. JSGT 11-4 (AC 65-9A)
When performing a magnetic particle inspection, defects that are perpendicular to the lines of flux are more easily detected because the flaw produces a large disruption in the magnetic field. Answer (B) is incorrect because defects that are parallel, not perpendicular, to the flux lines cause minimal disruptions in the magnetic field. Answer (C) is wrong because, if a flaw exists perpendicular to the magnetic field, a disruption will exist.

8236. E02
If dye penetrant inspection indications are not sharp and clear, the most probable cause is that the part

A — was not correctly degaussed before the developer was applied.
B — is not damaged.
C — was not thoroughly washed before developer was applied.

8236. Answer C. JSGT 11-4 (AC 65-9A)
It is very important to remove all excess penetrant from the surface of a part at the completion of the dwell time. Any remaining penetrant on the part may result in indications that are not clear, or are completely masked by the background indication.

8237. E02
(1) An aircraft part may be demagnetized by subjecting it to a magnetizing force from alternating current that is gradually reduced in strength.
(2) An aircraft part may be demagnetized by subjecting it to a magnetizing force from direct current that is alternately reversed in direction and gradually reduced in strength.
Regarding the above statements,

A — both No. 1 and No. 2 are true.
B — only No. 1 is true.
C — only No. 2 is true.

8238. E02
The pattern for an inclusion is a magnetic particle buildup forming

A — a fernlike pattern.
B — a single line.
C — parallel lines.

8239. E02
A part which is being prepared for dye penetrant inspection should be cleaned with

A — a volatile petroleum-base solvent.
B — the penetrant developer.
C — water-base solvents only.

8240. E02
Under magnetic particle inspection, a part will be identified as having a fatigue crack under which condition?

A — The discontinuity pattern is straight.
B — The discontinuity is found in a nonstressed area of the part.
C — The discontinuity is found in a highly stressed area of the part.

8241. E02
In performing a dye penetrant inspection, the developer

A — seeps into a surface crack to indicate the presence of a defect.
B — acts as a blotter to produce a visible indication.
C — thoroughly cleans the surface prior to inspection.

8237. Answer A. JSGT 11-12 (AC 65-9A)
By introducing a part to a decreasing, alternating magnetic field, the magnetic poles constantly change direction as their strength decreases. This is continued until the part is no longer magnetized. Reversing the polarity of direct current makes it functionally the same as alternating current.

8238. Answer C. JSGT 11-12 (AC 65-9A)
Inclusions are nonmetallic materials, such as slag, that have been trapped in a material. Large inclusions near the surface or open to the surface usually appear elongated or as parallel lines.

8239. Answer A. JSGT 11-5 (AC 65-9A)
The success of a penetrant inspection depends a great deal on the cleanliness of the part being tested. Precleaning with vapor degreaser or a volatile petroleum base solvent is a necessary step. Many penetrant kits provide an aerosol can of cleaning solvent to be used before and after testing. Water-base solvents (answer C) are typically not used for cleaning parts since they are not that effective on grease.

8240. Answer C. JSGT 11-11 (AC 65-9A)
Fatigue cracks are found in the highly stressed areas of a part, or where a stress concentration exists. Answer (A) is incorrect because, fatigue cracks offer clear, sharp patterns which are generally uniform and jagged in appearance.

8241. Answer B. JSGT 11-4 (AC 65-9A)
The principle action of the developer is to "blot" or draw the penetrant out of a defect making the defect visible. Answer (A) is incorrect because it is the penetrant, not the developer, that seeps into a surface crack to indicate the presence of a defect. Answer (C) is wrong because surface cleaning is left to a volatile petroleum-based solvent, not the developer.

8242.   E02
What defects will be detected by magnetizing a part using continuous longitudinal magnetization with a cable?

A — Defects perpendicular to the long axis of the part.
B — Defects parallel to the long axis of the part.
C — Defects parallel to the concentric circles of magnetic force within the part.

8243.   E02
Circular magnetization of a part can be used to detect which defects?

A — Defects parallel to the long axis of the part.
B — Defects perpendicular to the long axis of the part.
C — Defects perpendicular to the concentric circles of magnetic force within the part.

8244.   E02
(1) In nondestructive testing, a discontinuity may be defined as an interruption in the normal physical structure or configuration of a part.
(2) A discontinuity may or may not affect the usefulness of a part.

Regarding the above statements,

A — only No. 1 is true.
B — only No. 2 is true.
C — both No. 1 and No. 2 are true.

8278.   E05
(Refer to figure 44) Identify the weld caused by an excessive amount of acetylene.

A — 4.
B — 1.
C — 3.

8242. Answer A. JSGT 11-8 (AC 65-9A)
In longitudinal magnetization, a magnetic field is produced that is parallel to the axis of the part. This type of magnetization detects faults that are perpendicular to the axis of the part. Answers (B) and (C) are wrong because the lines of flux generated must intersect a defect at nearly a 90 degree angle.

8243. Answer A. JSGT 11-7 (AC 65-9A)
Circular magnetization is obtained by passing the current through the part, and indicates defects that are parallel to the axis of the part. Answers (B) and (C) are wrong because the magnetic flux lines must intersect a defect at nearly a 90 degree angle.

8244. Answer C. JSGT 11-7 (AC 65-9A)
A discontinuity may be defined as an interruption in the normal physical structure or configuration of a part such as a crack, seam, or inclusion. However, a discontinuity may or may not affect the usefulness of a part.

8278. Answer C. JSGT 11-4 (AC 65-9A)
When an excessive amount of acetylene is used, the puddle has a tendency to boil. This often leaves slight bumps along the center, and craters at the finish of the weld. Cross-checks are apparent if the body of the weld is sound. If the weld were cross-sectioned, pockets and porosity would be visible. This condition is illustrated in selection 3 (answer C).

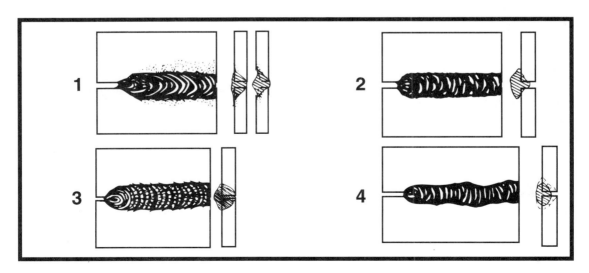

FIGURE 44.—Welds.

8279. E05
(Refer to figure 44) Select the illustration which depicts a cold weld.

A — 3.
B — 2.
C — 4.

8281. E05
Holes and a few projecting globules are found in a weld. What action should be taken?

A — Thoroughly clean the area and reweld over the first bead to fill gaps and obtain uniform strenght.
B — Remove all the old weld and reweld the joint.
C — Grind the rough surface smooth and reweld the joint.

8282. E05
Which condition indicates a part has cooled too quickly after being welded?

A — Cracking adjacent to the weld.
B — Discoloration of the base metal.
C — Gas pockets, porosity, and slag inclusions.

8283. E05
Select a characteristic of a good gas weld.

A — The depth of penetration shall be sufficient to ensure fusion of the filler rod.
B — The height of the weld bead should be 1/8 inch above the base metal.
C — The weld should taper off smoothly into the base metal.

8284. E05
One characteristic of a good weld is that no oxide should be formed on the base metal at a distance from the weld of more than

A — 1/2 inch.
B — 1 inch.
C — 1/4 inch.

8285. E05
(Refer to figure 45) What type weld is shown at A?

A — Fillet.
B — Butt.
C — Lap.

8286. E05
(Refer to figure 45) What type weld is shown at B?

A — Butt.
B — Double butt.
C — Fillet.

8279. Answer B. JSGT 11-4 (AC 65-9A)
A weld that is done with insufficient heat (a cold weld) appears rough and irregular and its edges will not feather into the base metal. This is illustrated in selection 2 (answer B).

8281. Answer B. JSGT 11-4 (AC 43.13-1A)
When checking the condition of a completed weld, there should be no signs of blowholes, porosity or projecting globules. If there is, the weld should be completely removed and the joint rewelded.

8282. Answer A. JSGT 11-4 (AC 65-9A)
Welding causes stresses to be set up in the material adjacent to the weld. In large parts, and in certain materials, these stresses can cause cracking if allowed to cool too rapidly.

8283. Answer C. JSGT 11-3 (AC 65-9A)
A good bead with proper penetration and good fusion is straight across the piece and has a smooth crowned surface that tapers off evenly into the base metal. The bead height should be 1/4 to 1/2 the thickness of the base metal.

8284. Answer A. JSGT 11-3 (AC 43.13-1A)
On a good weld, no oxide should be on the base metal more than 1/2 inch from the weld.

8285. Answer B. JSGT 11-3 (AC 65-9A)
Area A illustrates a butt weld. This type of weld is made by placing two pieces of material edge to edge, so that there is no overlap.

8286. Answer B. JSGT 11-3 (AC 65-9A)
Area B illustrates a double butt weld. On this type of weld a bead is applied on both sides of the joint. A single butt weld (answer A) is illustrated in area A.

Nondestructive Testing

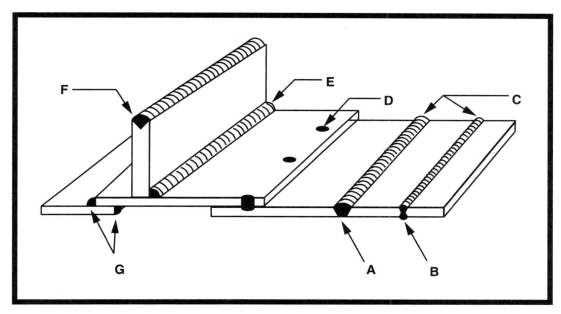

FIGURE 45.—Welds.

8287.   E05
(Refer to figure 45) What type of weld is shown at G?

A — Lap.
B — Butt.
C — Joint.

8287. Answer A. JSGT 11-3 (AC 65-15A)
Area G illustrates a lap weld. A lap weld is used when two pieces of material are overlapped and welded at the joint.

8288.   E05
On a fillet weld, the penetration requirement includes what percentage(s) of the base metal thickness?

A — 100 percent.
B — 25 to 50 percent.
C — 60 to 80 percent.

8288. Answer B. JSGT 11-4 (AC 65-9A)
A fillet weld is made along the joint of two metals that meet at a 90 degree angle. A fillet weld must penetrate 25 to 50 percent of the thickness of the base metal.

SECTION B
ELECTRONIC INSPECTIONS

Section B of Chapter 11 discusses the various types of electronic inspections currently used in aircraft maintenance. These methods include eddy current, ultrasonic, radiological, and pulse-echo inspections. The FAA Test questions taken from the information in this section include:

8221, 8222, 8224, 8227.

8221.   E01
Which of these nondestructive testing methods is suitable for the inspection of most metals, plastics, and ceramics for surface and subsurface defects?

A — Eddy current inspection.
B — Magnetic particle inspection.
C — Ultrasonic inspection.

8221. Answer C. JSGT 11-15 (AC 65-9A)
Only ultrasonic inspection meets all of the criteria listed in the question. Eddy current inspection, (answer A) is eliminated because it requires a material that conducts electricity. Magnetic particle inspection, (answer B) requires a ferromagnetic material and, therefore, will not work on plastic or ceramics.

8222. E01

What nondestructive testing method requires little or no part preparation, is used to detect surface or near-surface defects in most metals, and may also be used to separate metals or alloys and their heat-treat conditions?

A — Eddy current inspection.
B — Ultrasonic inspection.
C — Magnetic particle inspection.

8224. E01

How many of these factors are considered essential knowledge for x-ray exposure?

1. Processing of the film.
2. Material thickness and density.
3. Exposure distance and angle.
4. Film characteristics.

A — One.
B — Three.
C — Four.

8227. E01

A mechanic has completed a bonded honeycomb repair using the potted compound repair technique. What nondestructive testing method is used to determine the soundness of the repair after the repair has cured?

A — Eddy current test.
B — Metallic ring test.
C — Ultrasonic test.

8222. Answer A. JSGT 11-13 (AC 65-9A)

Eddy current inspection meets all the criteria of the question. Neither ultrasonic inspection (answer B) nor magnetic particle inspection (answer C) can distinguish between different alloys or heat-treat conditions.

8224. Answer B. JSGT 11-19 (AC 65-9A)

The factors affecting radiographic (x-ray) exposure include:

(a) material thickness and density,
(b) shape and size of the object,
(c) type of defect to be detected,
(d) characteristics of X-ray machine used,
(e) the exposure distance,
(f) the exposure angle,
(g) film characteristics, and
(h) types of intensifying screen, if used.

Three of these factors are listed in the question (answer B). Although film processing is an important procedure to ensure a valid inspection, it does not affect exposure.

8227. Answer B. JSGT 11-21 (AC 65-15A)

The metallic ring test is the simplest way to inspect honeycomb structures for delamination. When a coin (25-cent piece) is lightly bounced against a solid structure, a clear metallic ring is heard. If delamination is present a dull thud is heard. Eddy current (answer A) is incorrect because it requires the material being tested to accept an electric charge and composites are not conductive. Ultrasonic testing (answer C) is incorrect because ultrasound works only on solid structures and bonded honeycomb is not a solid structure.

CHAPTER 12

CLEANING AND CORROSION

SECTION A
AIRCRAFT CLEANING

Since most aircraft contain some metal, they require constant inspection and cleaning to minimize the destructive effects of corrosion. However, to prevent damaging an aircraft, specific procedures must be adhered to when cleaning various aircraft components. Section A of Chapter 12 looks at procedures and compounds used to clean aircraft. The following FAA Test questions are taken from this section:

    8347, 8348, 8350, 8352, 8353, 8354, 8355, 8368, 8369.

8347.   G01
A primary reason why ordinary or otherwise nonapproved cleaning compounds should not be used when washing aircraft is because their use can result in

A — hydrogen embrittlement in metal structures.
B — hydrogen embrittlement in nonmetallic materials.
C — a general inability to remove compound residues.

8347. Answer A. JSGT 12-2 (AC 43-4A)
Washing an aircraft with nonapproved cleaning compounds can cause a chemical reaction known as hydrogen embrittlement. Hydrogen embrittlement results when a chemical reaction produces hydrogen gas that is absorbed by the base metal. This process subsequently reduces a metal's ductility and causes the metal to become more brittle, allowing cracks and stress corrosion to occur.

8348.   G01
How may magnesium engine parts be cleaned?

A — Soak in a 20 percent caustic soda solution.
B — Spray with MEK (methyl ethyl ketone).
C — Wash with a commercial solvent, decarbonize, and scrape or grit blast.

8348. Answer C. JSGT 12-3 (AC 65-12A)
Magnesium engine parts may be cleaned using commercial solvents, scrapers, and grit blasting. However, extreme care must be taken to make certain that the chemical cleaner used is safe for use on magnesium.

8350.   G01
Select the solvent recommended for wipedown of cleaned surfaces just before painting.

A — Aliphatic naphtha.
B — Dry-cleaning solvent.
C — Aromatic naphtha.

8350. Answer A. JSGT 12-3 (AC 65-9A)
Aliphatic naphtha is the solvent recommended for wipedown on cleaned surfaces just before painting. In addition to metals, aliphatic naphtha is also safe to use on acrylic plastics and rubber.

8352.   G01
What is used for general cleaning of aluminum surfaces by mechanical means?

A — Carborundum paper.
B — Aluminum wool.
C — Crocus cloth.

8352. Answer B. JSGT 12-3 (AC 65-9A)
Aluminum wool is typically used for general cleaning of aluminum surfaces. It is important not to introduce material to the surface of the aluminum that may result in dissimilar metal corrosion. For this reason, steel wool or any abrasive product containing metals other than aluminum are not satisfactory.

8353. G01
Select the solvent used to clean acrylics and rubber.

A — Aliphatic naphtha.
B — Methyl ethyl ketone.
C — Aromatic naphtha.

8354. G01
Fayed surfaces cause concern in chemical cleaning because of the danger of

A — forming passive oxides.
B — entrapping corrosive materials.
C — corrosion by imbedded iron oxide.

8355. G01
Caustic cleaning products used on aluminum structures have the effect of producing

A — passive oxidation.
B — improved corrosion resistance.
C — corrosion.

8368. G02
Why is a plastic surface flushed with fresh water before it is cleaned with soap and water?

A — To prevent crazing.
B — To prevent scratching.
C — To prevent discoloration and embrittlement.

8369. G02
What should be done to prevent rapid deterioration when oil or grease come in contact with a tire?

A — Wipe the tire thoroughly with a dry cloth, and then rinse with clean water.
B — Wipe the tire with a dry cloth followed by a wash down and rinse with soap and water.
C — Wipe the tire with a cloth dampened with aeromatic naphtha and then wipe dry with a clean cloth.

8353. Answer A. JSGT 12-3 (AC 65-9A)
Aliphatic naphtha, which is typically used to wipe down surfaces prior to painting, is also used to clean acrylics and rubber.

8354. Answer B. JSGT 12-3 (AC 65-9A)
Faying surfaces exist at joints that are fitted or joined tightly, as in a riveted seam. When using chemical cleaners around fayed surfaces, corrosive chemical residue can get into the small openings of the fayed surface. Because of this, you must ensure that all chemical residue is removed and/or neutralized. Any residue that becomes entrapped can become a corrosion problem.

8355. Answer C. JSGT 12-3 (AC 65-9A)
The term caustic means having the ability to burn, corrode, or eat away. By definition these are cleaners that can produce corrosion.

8368. Answer B. JSGT 12-3 (AC 65-9A)
Before applying soap and water to plastic surfaces, flush the plastic with fresh water to dissolve salt deposits and wash away dust particles. This helps prevent scratching when the surface is washed.

8369. Answer B. JSGT 12-3 (AC 65-9A)
Lubricating oil causes deterioration of rubber tires and should be removed as soon as it is detected. To do this, wipe the oil from the tire with a dry cloth and wash the tire with soap and water.

Cleaning and Corrosion

SECTION B
TYPES OF CORROSION

Aircraft are subject to attack by a number of different kinds of corrosion. In fact, as aircraft age, the prevention of corrosion becomes a primary concern. Section B of Chapter 12 discusses the different types of corrosion found on aircraft. FAA Test questions taken from this section include:

8245, 8356, 8357, 8359, 8360, 8363, 8365, 8371, 8372, 8374, 8375, 8376, 8377, 8378.

8245. E03
What type of corrosion attacks grain boundaries of aluminum alloys which are improperly or inadequately heat treated?

A — Filiform.
B — Intergranular.
C — Fretting.

8245. Answer B. JSGT 12-12 (AC 65-9A)
Intergranular corrosion is a type of dissimilar metal corrosion along the grain boundaries of aluminum that has been improperly heat treated.

8356. G02
Fretting corrosion is most likely to occur

A — when two surfaces fit tightly together but can move relative to one another.
B — only when two dissimilar metals are in contact.
C — when two surfaces fit loosely together and can move relative to one another.

8356. Answer A. JSGT 12-13 (AC 65-9A)
Fretting corrosion occurs when two surfaces are held tightly together, but are still subject to a small amount of relative movement. When the movement between the parts is small, the debris created by the rubbing action remains between the surfaces and acts as an abrasive to cause further erosion.

8357. G02
The rust or corrosion that occurs with most metals is the result of

A — a tendency for them to return to their natural state.
B — blocking the flow of electrons in homogenous metals, or between dissimilar metals.
C — electron flow in or between metals from cathodic to anodic areas.

8357. Answer A. JSGT 12-5 (AC 43-4A)
Corrosion occurs because of the tendency of metals to return to their natural state. Noble metals such as gold and platinum do not corrode because they are chemically uncombined in their natural state.

8359. G02
Which of the listed conditions is NOT one of the requirements for corrosion to occur?

A — The presence of an electrolyte.
B — Electrical contact between an anodic area and cathodic area.
C — The presence of a passive oxide film.

8359. Answer C. JSGT 12-6 (AC 43-4A)
Four conditions must exist before corrosion can occur: (1) Presence of a metal that will corrode (anode), (2) Presence of a dissimilar metal (cathode), (3) Presence of an electrolyte, and (4) Electrical contact between anode and cathode. The presence of a passive oxide film (answer C) is not necessary for corrosion to occur.

8360. G02
The lifting or flaking of the metal at the surface due to delamination of grain boundaries caused by the pressure of corrosion residual product buildup is called

A — brinelling.
B — granulation.
C — exfoliation.

8360. Answer C. JSGT 12-12 (AC 65-9A)
The pressure of corrosion residual product buildup can cause delamination of grain boundaries resulting in lifting or flaking, and is called exfoliation.

8363.   G02
Intergranular corrosion in aluminum alloy parts

A — may be detected by surface pitting, and white, powdery deposit formed on the surface of the metal.
B — commonly appears as threadlike filaments of corrosion products under a dense film of paint.
C — cannot always be detected by surface indications.

8365.   G02
A primary cause of intergranular corrosion is

A — improper heat treatment.
B — dissimilar metal contact.
C — improper application of primer.

8371.   G02
Corrosion caused by galvanic action is the result of

A — excessive anodization.
B — contact between two unlike metals.
C — excessive etching.

8372.   G02
Which of these materials is the most anodic?

A — Cadmium.
B — 7075-T6 aluminum alloy.
C — Magnesium.

8374.   G02
Which of these materials is the most cathodic?

A — Zinc.
B — 2024 aluminum alloy.
C — Stainless steel.

8375.   G02
Galvanic corrosion is likely to be most rapid and severe when

A — the surface area of the cathodic metal is smaller than surface area of the anodic metal.
B — the surface areas of the anodic and cathodic metals are approximately the same.
C — the surface area of the anodic metal is smaller than the surface area of the cathodic metal.

8376.   G02
One way of obtaining increased resistance to stress corrosion cracking is by

A — relieving compressive stresses on the metal surface.
B — creating compressive stresses on the metal surface.
C — producing nonuniform deformation while cold working during the manufacturing process.

8363. Answer C. JSGT 12-12 (AC 65-9A)
Intergranular corrosion takes place within the grain structure of the metal and, therefore, cannot always be detected by surface indications. It is very difficult to detect visually in its early stages. The best way to detect intergranular corrosion is with ultrasonic and eddy current inspection methods.

8365. Answer A. JSGT 12-12 (AC 65-9A)
The primary cause of intergranular corrosion is improper heat treatment. For example, certain aluminum alloys must be transferred from the heat-treat furnace to the quench in a very short time (as little as 10 seconds). If quenching is delayed, the grain structure of an alloy can become irregular and cause the metal to lift or flake.

8371. Answer B. JSGT 12-9 (AC 65-9A)
Contact of different bare metals creates an electrolytic action when moisture is present. This is sometimes referred to as galvanic corrosion. Electrolyte and dissimilar metal corrosion are additional terms that may be used.

8372. Answer C. JSGT 12-6 (AC 65-9A)
Magnesium is the most anodic metal on the Electrochemical Series of metals chart. In other words, magnesium corrodes very easily.

8374. Answer C. JSGT 12-6 (AC 43-4A)
Stainless steel is very cathodic, meaning it does not corrode easily.

8375. Answer C. JSGT 12-9 (AC 43-4A)
Galvanic corrosion occurs when two dissimilar metals make electrical contact in the presence of an electrolyte. If the surface area of the anode (the corrosive metal) is smaller than the cathode, corrosion will be rapid and severe.

8376. Answer B. JSGT 12-13 (AC 43.13-1A)
Shot peening metal surfaces increases their resistance to stress-corrosion cracking by creating compressive stresses on the surface. However, the compressive stresses should be overcome by applied tensile stress before the surface is exposed to any tension load.

Cleaning and Corrosion

8377.   G02
(1) In the corrosion process, it is the cathodic area or dissimilar cathodic material that corrodes.
(2) In the Galvanic or Electro-Chemical Series for metals, the most anodic metals are those that will give up electrons most easily.
Regarding the above statements,

A — only No. 1 is true.
B — only No. 2 is true.
C — both No. 1 and No. 2 are true.

8377. Answer B. JSGT 12-5 (AC 43-4A/AC 65-9A)
Only statement number 2 (answer B) is true. In the galvanic series of metals, the anodic metals give up electrons most easily. Therefore, the anode metal corrodes.

8378.   G02
Spilled mercury on aluminum

A — increases susceptibility to hydrogen embrittlement.
B — may cause impaired corrosion resistance if left in prolonged contact.
C — causes rapid and severe corrosion that is very difficult to control.

8378. Answer C. JSGT 12-14 (AC 43-4A)
Mercury attacks aluminum through a process called amalgamation. It causes pitting and intergranular corrosion and is very difficult to control.

SECTION C
CORROSION DETECTION

Section C of Chapter 12 presents methods of detecting corrosion once it has become established in an aircraft structure. In addition, several of the common areas where corrosion typically occurs are discussed. There are no FAA Test questions taken from the information presented in this section.

SECTION D
TREATMENT OF CORROSION

Once corrosion is found on an aircraft structure it must be removed. Furthermore, the corroded area must be protected from further attack. Chapter 12 Section D discusses the various methods of corrosion removal and control used in aviation. The FAA Test questions drawn from this section include:

  8349, 8358, 8361, 8364, 8366, 8367, 8370, 8373.

8349.   G01
When an anodized surface coating is damaged in service, it can be partially restored by

A — use of a metal polish.
B — chemical surface treatment.
C — a suitable mild cleaner.

8349. Answer B. JSGT 12-26 (AC 65-9A)
Anodizing is a common surface treatment of aluminum alloys. It is an electrolytic process requiring special equipment and is not generally done in the repair shop. When this coating is damaged in service, it can only be partially restored by chemical surface treatment.

8358.   G02
Which of the following are the desired effects of using Alodine on aluminum alloy?

1. A slightly rough surface.   A—3 and 4.
2. Relieved surface stresses.   B—1, 2, and 4.
3. A smooth painting surface.   C—1 and 4.
4. Increased corrosion resistance..

8358. Answer C. JSGT 12-26 (AC 65-9A)
Alodizing is a simple chemical treatment for all aluminum alloys to increase their corrosion resistance and to improve paint-bonding qualities.

8361. G02
A nonelectrolytic chemical treatment for aluminum alloys to increase corrosion resistance and paint-bonding qualities is called

A — anodizing.
B — alodizing.
C — dichromating.

8364. G02
What may be used to remove corrosion from highly stressed steel surfaces?

A — Steel wire brushes.
B — Fine-grit aluminum oxide.
C — Medium-grit carborundum paper.

8366. G02
Corrosion should be removed from magnesium parts with a

A — silicon carbide brush.
B — carborundum abrasive.
C — stiff, hog-bristle brush.

8367. G02
Why is it important not to rotate the crankshaft after the corrosion preventive mixture has been put into the cylinders on engines prepared for storage?

A — Fuel may be drawn into one or more cylinders and dilute or wash off the corrosion preventive mixture.
B — The seal of corrosion preventive mixture will be broken.
C — The link rods may be damaged by hydraulic lock.

8370. G02
Galvanic action caused by dissimilar metal contact may best be prevented by

A — placing a nonporous dielectric material between the surfaces.
B — cleaning both surfaces with a non-residual solvent.
C — application of paper tape between the surfaces.

8373. G02
The interior surface of sealed structural steel tubing wouild be best protected against corrosion by which of the following?

A — Charging the tubing with dry nitrogen prior to sealing.
B — evacuating the tubing before sealing.
C — a coating of hot linseed oil.

8361. Answer B. JSGT 12-26 (AC 65-9A)
Alodizing is a chemical treatment for corrosion on aluminum alloys and is often used to increase paint-bonding properties. Anodizing (answer A) is an electrolytic treatment that prevents corrosion.

8364. Answer B. JSGT 12-27 (AC 65-9A)
Corrosion on highly stressed steel should be removed using mild abrasive papers such as rouge, fine grit aluminum oxide, or even pumice. The reason for this is that highly abrasive materials can cause surface scratches or changes in the surface structure from overheating.

8366. Answer C. JSGT 12-29 (AC 65-9A)
Magnesium is a highly-active metal, and contact with most other metals will result in dissimilar metal corrosion. To avoid the danger of residue from metal abrasive or wire brushes, a stiff hog-bristle brush is typically used to remove corrosion from magnesium.

8367. Answer B. JSGT 12-30 (AC 65-12A)
When preparing an engine for long term storage, a corrosion preventive mixture is sprayed into the cylinders to form a seal against oxidation. After this mixture is applied, the propeller shaft should not be moved. Any movement causes the pistons to move which breaks the seal of the corrosion preventive mixture.

8370. Answer A. JSGT 12-30 (AC 65-9A)
To prevent corrosion when dissimilar metals are joined together, a protective separator must be employed. The separating material depends on the metals used and may be a light coat of primer, aluminum tape, washers, grease, or sealant.

8373. Answer C. JSGT 12-21 (AC 43.13-1A)
A small amount of water entrapped in a tube can corrode through the tubing in a very short time. To protect tubing, hot linseed oil is used to coat the interior surfaces.

CHAPTER 13

GROUND HANDLING AND SERVICING

SECTION A
SHOP SAFETY

Section A of Chapter 13 discusses safety in the shop. Included is information regarding fire safety, the importance of protective clothing, including safety goggles, and the safety precautions that should be taken when jacking or hoisting an aircraft. While shop safety is critical, no FAA Test questions are taken from this subject area.

SECTION B
FLIGHT LINE SAFETY

Chapter 13, Section B discusses the hazards of working around aircraft on the flight line. Flight line safety is especially critical because of the risks ground personnel are exposed to in the vicinity of spinning propellers and idling turbine engines. The following FAA Test questions are taken from Section B:

8308, 8309, 8310, 8311, 8312, 8313, 8314, 8315, 8316, 8317, 8318, 8319, 8320, 8321, 8322, 8323, 8326, 8327, 8328, 8329, 8330, 8331, 8332, 8333, 8334.

8308.   **F01**
During starting of a turbine powerplant using a compressed air starter, a hung start occurred. Select the proper procedure.

A — Shut the engine down
B — Re-engage the starter
C — Advance power level to increase RPM

8309.   **F01**
A hung start in a jet engine is often caused by

A — malfunctions in the ignition system.
B — the starter cutting off too soon.
C — an excessively rich fuel/air mixture.

8310.   **F01**
Which statement below reflects a typical requirement when towing some aircraft?

A — Discharge all hydraulic pressure to prevent accidental operation of the nosewheel steering mechanism.
B — Tailwheel aircraft must be towed backwards.
C — If the aircraft has a steerable nosewheel, the torque-link lock should be set to full swivel.

8308. Answer A. JSGT 13-20 (AC 65-9A)
A hung start is one in which the engine starts but does not accelerate to normal starting RPM. If a hung start occurs, the engine should be shut down.

8309. Answer B. JSGT 13-20 (AC 65-9A)
Hung starts are often the result of insufficient power to the starter, or the starter cutting off before the engine starts self-accelerating. To help prevent a hung start, many operators insist on ground power assistance when starting.

8310. Answer C. JSGT 13-23 (AC 65-9A)
Ground handling should be done according to the aircraft manufacturer's instructions. When towing an aircraft with a steerable nosewheel, you should verify that the locking scissors are set to full swivel. Failure to follow the proper procedures can result in damage to the aircraft.

8311. F01
Which statement(s) is/are true regarding tiedown of small aircraft?

1. Manila (hemp) rope has a tendency to stretch when it gets wet.
2. Nylon or dacron rope is preferred to manila rope.
3. The aircraft should be headed downwind in order to eliminate or minimize wing lift.
4. Leave the nosewheel or tailwheel unlocked.

A — 1, 2, 3, and 4.
B — 1 and 2.
C — 2.

8311. Answer C. JSGT 13-14 (AC 65-9A)
When tieing down an aircraft, nylon or Dacron rope is preferred because it won't shrink, mildew, or rot.

8312. F01
When approaching the front of an idling jet engine, the hazard area extends forward of the engine approximately

A — 10 feet.
B — 15 feet.
C — 25 feet.

8312. Answer C. JSGT 13-14 (AC 65-9A)
Whether at idle or at full thrust, the hazard area in front of a jet engine extends out 25 feet. In the rear, the hazard area may extend from 100 to 200 feet, depending on the power setting.

8313. F01
Which of the following is the most satisfactory extinguishing agent for use on a carburetor or intake fire?

A — Dry chemical.
B — A fine, water mist.
C — Carbon dioxide.

8313. Answer C. JSGT 13-18 (AC 65-9A)
Carbon dioxide is the most satisfactory extinguishing agent for carburetor or intake fires. CO_2 is preferred, because it leaves no residue.

8314. F01
(Refer to figure 50) Identify the signal to engage rotor on a rotorcraft.

A — 1.
B — 3.
C — 2.

8314. Answer B. JSGT 13-22 (AC 65-9A)
Use of standard hand signals is important for communication in high noise areas. Pointing one hand at the rotor and moving the other hand in a circle, indicates that it is safe to engage the rotor.

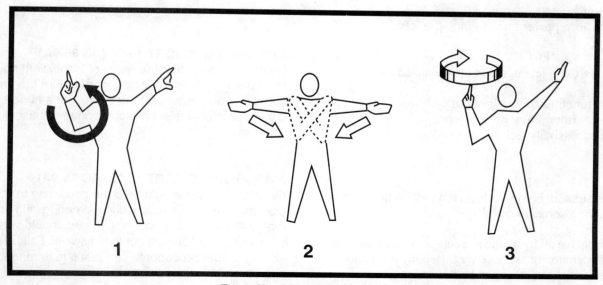

Figure 50.—Marshalling Signals.

Ground Handling and Servicing

8315.   F01
If a radial engine has been shut down for more than 30 minutes, the propeller should be rotated through at least two revolutions to

A — check for hydraulic lock.
B — check for leaks.
C — prime the engine.

8316.   F01
The priming of a fuel injected horizontally opposed engine is accomplished by placing the fuel control lever in the

A — IDLE-CUTOFF position.
B — AUTO-RICH position.
C — FULL-RICH position.

8317.   F01
The most important condition to be monitored during start after fuel flow begins in a turbine engine is the

A — EGT, TIT, or ITT.
B — RPM.
C — oil pressure.

8318.   F01
How is a flooded engine, equipped with a float-type carburetor, cleared of excessive fuel?

A — Crank the engine with the starter or by hand, with the mixture control in cutoff, ignition switch off, and the throttle fully open, until the fuel charge has been cleared.
B — Turn off the fuel and the ignition. Discontinue the starting attempt until the excess fuel has cleared.
C — Crank the engine with the starter or by hand, with the mixture control in cutoff, ignition switch on, and the throttle fully open, until the excess fuel has cleared or until the engine starts.

8315. Answer A. JSGT 13-19 (AC 65-9A)
Oil seeping past the piston rings or incomplete scavenging of oil in the cylinders of a radial engine may result in an accumulation of oil in the lower cylinders. Before starting a radial engine that has been shut down for more than 30 minutes, you should check for hydraulic lock by turning the propeller three or four complete revolutions. If liquid is present in the cylinders it is indicated by a resistance to turn, or by the prop stopping abruptly. All liquid must be cleared from the cylinders before attempting to start the engine.

8316. Answer C. JSGT 13-19 (AC 65-9A)
To prime most fuel injection systems installed on horizontally opposed engines, you place the mixture control in the full-rich position. If the fuel control lever is placed in the idle-cutoff position (answer A) no fuel can flow to the engine and there is no such thing as an auto-rich position (answer B) on a piston engine.

8317. Answer A. JSGT 13-20 (AC 65-9A)
Light-off of a turbine engine is indicated by an increase in exhaust gas temperature (EGT), turbine inlet temperature (TIT), and interstage and turbine temperature (ITT). These temperatures should be monitored closely during start to prevent engine damage. In a piston engine, oil pressure (answer C) is the most important condition to monitor.

8318. Answer A. JSGT 13-19 (AC 65-9A)
The best way to clear a flooded engine is to crank the engine with the mixture in the cutoff position, ignition turned off, and the throttle fully open. This purges the excess fuel from the cylinders.

8319.   F01
(Refer to figure 51) Which marshalling signal should be given if a taxiing aircraft is in danger of striking an object?

A — 1.
B — 2.
C — 3.

8319. Answer C. JSGT 13-21 (AC 65-9A)
Anyone working around moving aircraft should know the standard FAA hand signals. Hands over the head and moving from shoulder width to crisscross is the signal for an emergency stop (answer C).

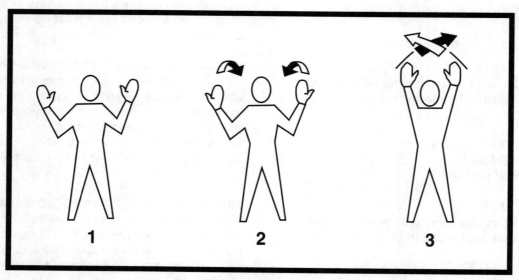

Figure 51.—Marshalling Signals.

8320.   F01
Generally, when an induction fire occurs during starting of a reciprocating engine, the first course of action should be to

A — direct carbon dioxide into the air intake of the engine.
B — continue cranking and start the engine if possible.
C — close the throttle.

8320. Answer B. JSGT 13-19 (AC 65-9A)
Induction fires during starting are surprisingly common and you should be ready to deal with them at any time. If a fire occurs during the start, you should continue cranking and attempt to start the engine. This draws the fire into the engine. If the engine does not start and the fire continues, discontinue the start attempt. A carbon dioxide fire extinguisher may then be discharged into the engine's intake (answer A).

8321.   F01
When starting and ground operating an aircraft's engine, the aircraft should be positioned to head into the wind primarily

A — to aid in achieving and maintaining the proper air and fuel mixture flow in the engine.
B — for engine cooling purposes.
C — to help cancel out engine torque effect.

8321. Answer B. JSGT 13-19 (AC 65-9A)
A headwind will aid in engine cooling during ground operations.

8322.   F01
When approaching the rear of an idling turbojet engine, the hazard area extends aft of the engine approximately

A — 200 feet.
B — 100 feet.
C — 50 feet.

8322. Answer B. JSGT 13-14 (AC 65-9A)
An idling turbojet engine creates high velocity gases that extend 100 feet aft of the engine. At takeoff power, the hazard area extends back 200 feet (answer A).

Ground Handling and Servicing

8323. F01
During starting of a turbojet powerplant using a compressed air starter, a hot start occurrence was recorded. Select what happened from the following.

A — The pneumatic starting unit overheated.
B — The powerplant was preheated before starting.
C — The fuel/air mixture was excessively rich.

8326. F01
During towing operations

A — a person should be in the cockpit to watch for obstructions.
B — persons should be stationed at the nose, each wingtip, and the empennage at all times.
C — a qualified person should be in the cockpit to operate brakes.

8327. F01
The tendency of tailwheel-type airplanes to weathervane is greatest while taxiing with a

A — direct crosswind.
B — quartering headwind.
C — tailwind.

8328. F01
A tailwheel-type airplane has a greater tendency to weathervane during taxi than a nosewheel-type because on a tailwheel airplane, the

A — vertical stabilizer to fuselage proportion is greater.
B — surface area ratio behind the pivot point (main gear) is greater.
C — surface area ratio behind the pivot point (main gear) is less.

8329. F01
When taxiing (or towing) an aircraft, a flashing red light from the control tower means

A — stop and wait for a green light.
B — move clear of the runway/taxiway immediately.
C — return to starting point.

8330. F01
A person should approach or leave a helicopter in the pilot's field of vision whenever the engine is running in order to avoid

A — the tail rotor.
B — the main rotor.
C — blowing dust or debris caused by rotor downwash.

8323. Answer C. JSGT 13-20 (AC 65-9A)
A hot start is characterized by excessive turbine inlet or exhaust gas temperatures. This condition is the result of an excessively RICH fuel/air mixture.

8326. Answer C. JSGT 13-23 (AC 65-9A)
When towing an aircraft, a qualified person should be in the cockpit to operate the aircraft's brakes, because the brakes on most tow vehicles are insufficient to stop the momentum of a large aircraft. Answers (A) and (B) are incorrect because a person does not have to be in the cockpit or stationed at the empennage when moving an aircraft.

8327. Answer A. JSGT 13-23 (AC 61-21A)
Because the center of gravity is located behind the main gear on tailwheel aircraft, they have an increased tendency to weathervane when taxiing in a direct crosswind.

8328. Answer B. JSGT 13-23 (AC 61-21A)
A tailwheel aircraft has a greater surface area behind the main gear (pivot point) than a nosewheel-type aircraft. Because of this, a tailwheel aircraft has a greater tendency to weathervane.

8329. Answer B. JSGT 13-23 (AC 65-9A)
A flashing red light from the control tower indicates that you should taxi clear of the runway/taxiway immediately. A steady red light means that you should stop (answer A) and a flashing white light means you should return to the starting point (answer C).

8330. Answer A. JSGT 13-14 (AC 91-32A)
When approaching a helicopter, you should be aware of propellers, rotors, and jet engine intakes and exhausts. One way to help ensure you remain clear of the tail rotor on a helicopter is to approach and leave the helicopter in the pilot's field of vision.

8331. F01

When taxiing (or towing) an aircraft, a flashing white light from the control tower means

A — move clear of the runway/taxiway immediately.
B — OK to proceed but use extreme caution.
C — return to starting point.

8331. Answer C. JSGT 13-23 (AC 65-9A)
A flashing white light from the control tower is a signal for you to return to the place you started from. An alternating red and green light indicates you should proceed with extreme caution (answer B), whereas, a flashing red light indicates you should move clear of the runway or taxiway you are on (answer A).

8332. F01

When taxiing (or towing) an aircraft, an alternating red and green light from the control tower means

A — move clear of the runway/taxiway immediately.
B — OK to proceed but use extreme caution.
C — return to starting point.

8332. Answer B. JSGT 13-23 (AC 65-9A)
Alternating red and green flashes from the tower indicate that it's O.K. to proceed but use extreme caution. A flahsing red light indicates you should move clear of the runway or taxiway immediately (answer A), and a flashing white light indicates you should return to where you started (answer C).

8333. F01

When parking a nosewheel-type airplane after taxiing (or towing), the nosewheel should be left

A — unlocked.
B — turned at a small angle.
C — pointed straight ahead.

8333. Answer C. JSGT 13-16 (AC 65-9A)
When parking a nose-wheel type aircraft, the nose-wheel should be left in the straight ahead position so the aircraft cannot move from side-to-side in the wind.

8334. F01

When first starting to move an aircraft while taxiing, it is important to

A — test the brakes.
B — closely monitor the instruments.
C — notify the control tower.

8334. Answer A. JSGT 13-23 (AC 61-21A)
After the engine is started and immediately after the aircraft begins moving, the brakes should be tested. If braking is unsatisfactory, the engine should be shut down immediately.

SECTION C
SERVICING AIRCRAFT

Section C of Chapter 13 discusses the maintenance practices used when servicing aircraft. Included is information on fueling and defueling, as well as things to watch for when operating ground power units (GPUs). The FAA Test questions based on this section include:

8324, 8325, 8335, 8336, 8337, 8338, 8339, 8340, 8341, 8342, 8343, 8344, 8345, 8346.

8324. F01

What effect will aviation gasoline mixed with jet fuel have on turbine powerplant efficiency?

A — No appreciable effect.
B — The tetraethyl lead in the gasoline forms deposits on the turbine blades.
C — The tetraethyl lead in the gasoline forms deposits on the compressor blades.

8324. Answer B. JSGT 13-28 (AC 65-9A)
When turbine engines are operated on a mixture of jet fuel and aviation gasoline, the TEL (lead) in the gasoline forms deposits on the turbine blades and vanes. Continuous use of mixed fuel typically causes a loss in engine efficiency. Answer (C) is incorrect because the exhaust gases containing the tetraethyl lead do not pass through the compressor blades.

Ground Handling and Servicing

8325. F01
(1) Jet fuel is of higher viscosity than aviation gasoline and therefore holds contaminants better.
(2) Viscosity has no relation to contamination of fuel.
Regarding the above statements,

A — only No. 1 is true.
B — both No. 1 and No. 2 are true.
C — neither No. 1 nor No. 2 is true.

8325. Answer A. JSGT 13-31 (AC 65-9A)
The higher the viscosity of a fuel, the greater its ability to hold contaminants in suspension. Jet fuel does have a high viscosity and, therefore, is more susceptible to contamination than aviation gasoline. Only statement number 1 is true.

8335. F02
The color of 100LL fuel is

A — blue.
B — colorless or straw.
C — red.

8335. Answer A. JSGT 13-28 (AC 65-9A)
100LL aviation fuel is colored blue (answer A), jet A is colorless or may have a straw tint (answer B), and 80/87 is dyed red (answer C).

8336. F02
How are aviation fuels, which possess greater anti-knock qualities than 100 octane, classified?

A — According to the milliliters of lead.
B — By reference to normal heptane.
C — By performance numbers.

8336. Answer C. JSGT 13-28 (AC 65-9A)
Octane and performance numbers designate the anti-knock value of a fuel. The higher the grade, the more compression the fuel can withstand without detonating. The octane scale goes to 100, which represents the anti-knock characteristics of pure iso-octane. For fuels with higher anti-knock characteristics, performance numbers are used.

8337. F02
Why is ethylene dibromide added to aviation gasoline?

A — To remove zinc silicate deposits from the spark plugs.
B — To scavenge lead oxide from the cylinder combustion chambers.
C — To increase the anti-knock rating of the fuel.

8337. Answer B. JSGT 13-28 (AC 65-9A)
Ethylene dibromide is added to aviation gasoline to scavenge lead oxide from the cylinders. The ethylene dibromide combines with the lead oxide in aviation fuels to form a lead bromide that passes out the exhaust.

8338. F02
Both gasoline and kerosene have certain advantages for use as turbine fuel. Which statement is true in reference to the advantages of each?

A — Kerosene has a higher heat energy per unit weight than gasoline.
B — Gasoline has a higher heat energy per unit volume than kerosene.
C — Kerosene has a higher heat energy per unit volume than gasoline.

8338. Answer C. JSGT 13-26 (AST Spec. D-910, D-1655)
Aviation gasoline (100LL) has a heat energy of 18,720 BTU/lb. Jet A fuel (kerosene) has a heat energy of 18,439 BTU/lb. Since Jet A weighs 6.7 lbs/gal, its heat energy is 123,541 BTU/gal. 100LL weighs 6 lbs/gal and, therefore, has a heat energy of 112,320 BTU/gal. Therefore, kerosene has a higher heat energy per unit volume (answer C).

8339. F02
What must accompany fuel vaporization?

A — An absorption of heat.
B — A decrease in vapor pressure.
C — A reduction in volume.

8339. Answer A. JSGT 13-27 (AC 65-9A)
When any material, including fuel, changes from a liquid to a vapor state, it must absorb heat from its surroundings.

8340.   F02
Characteristics of detonation are

A — cylinder pressure remains the same, excessive cylinder head temperature, and a decrease in engine power.
B — rapid rise in cylinder pressure, excessive cylinder head temperature, and a decrease in engine power.
C — rapid rise in cylinder pressure, cylinder head temperature normal, and a decrease in engine power.

8341.   F02
A fuel that vaporizes too readily may cause

A — hard starting.
B — detonation.
C — vapor lock.

8342.   F02
Jet fuel number identifiers are

A — performance numbers to designate the volatility of the fuel.
B — performance numbers and are relative to the fuel's performance in the aircraft engine.
C — type numbers and have no relation to the fuel's performance in the aircraft engine.

8343.   F02
The main differences between grades 100 and 100LL fuel are

A — volatility and lead content.
B — volatility, lead content, and color.
C — lead content and color.

8344.   F02
Characteristics of aviation gasoline are

A — high heat value, high volatility.
B — high heat value, low volatility.
C — low heat value, low volatility.

8345.   F02
Tetraethyl lead is added to aviation gasoline to

A — retard the formation of corrosives.
B — improve the gasoline's performance in the engine.
C — dissolve the moisture in the gasoline.

8346.   F02
A fuel that does not vaporize readily enough can cause

A — vapor lock.
B — detonation.
C — hard starting.

8340. Answer B. JSGT 13-27 (AC 65-9A)
Detonation is also known as engine knock and is caused by an abrupt explosion of the fuel/air mixture in the cylinder. This causes a rapid rise in cylinder pressures, an increase in cylinder head temperature, and a decrease in power. Detonation can cause damage to the cylinder head and/or piston.

8341. Answer C. JSGT 13-27 (AC 65-9A)
Volatility is a measure of the tendency of a liquid to vaporize under given conditions. If gasoline vaporizes too readily, vapor lock may occur (answer C). If it does not vaporize readily, hard starting may result (answer A).

8342. Answer C. JSGT 13-29 (AC 65-9A)
The numbers and letters used to identify different types of jet fuel have no relation to the fuel's rating or performance in the engine.

8343. Answer C. JSGT 13-28 (AC 65-9A)
The primary differences between 100 and 100LL fuel are the color and lead content. 100 grade aviation fuel (old designation 100/130) is green in color and contains more tetraethyl lead than the newer 100LL which is dyed blue.

8344. Answer A. JSGT 13-26 (AC 65-9A)
Aviation gasoline has a high heat value (approximately 18,720 BTUs per pound) and is highly volatile (5.5 to 7 PSI at 100F).

8345. Answer B. JSGT 13-28 (AC 65-9A)
Tetraethyl lead (TEL) is added to gasoline to improve the performance (anti-knock rating). The addition of TEL allows the engine to develop more power without knocking.

8346. Answer C. JSGT 13-27 (AC 65-9A)
A fuel that does not vaporize readily can cause hard starting (answer C); whereas, a fuel that vaporizes too readily can cause vapor lock (answer A).

CHAPTER 14

MAINTENANCE PUBLICATIONS, FORMS, AND RECORDS

SECTION A
MAINTENANCE PUBLICATIONS

Having and using the proper maintenance publications are every bit as important as having the correct tools for a given job. Chapter 14, Section A presents information on government publications such as Airworthiness Directives, Advisory Circulars, Type Certificate Data Sheets, and manufacturer's manuals. The FAA Test questions drawn from this section include:

8159, 8446, 8447, 8449, 8460, 8461, 8492, 8493, 8494, 8495, 8496, 8497, 8498, 8499, 8500, 8501, 8502, 8503, 8504, 8505, 8506, 8507, 8508, 8509, 8510, 8511, 8515, 8516, 8517, 8518, 8520, 8521, 8522, 8527, 8533, 8534, 8535.

8159.   **C01**
What FAA-approved document gives the leveling means to be used when weighing an aircraft?

A — Type Certificate Data Sheet.
B — AC 43.13-1A.
C — Manufacturer's maintenance manual.

8446.   **I01**
What is the means by which the FAA notifies aircraft owners and other interested persons of unsafe conditions and prescribes the condition under which the product may continue to be operated?

A — Airworthiness Directives.
B — Airworthiness Alerts.
C — Aviation Safety Data.

8447.   **I01**
Which is an appliance major repair?

A — Overhaul of a hydraulic pressure pump.
B — Repairs to a propeller governor or its control.
C — Troubleshooting and repairing broken circuits in landing lights circuits.

8159. Answer C. JSGT 14-8 (AC 65-9A)
The FAA-approved leveling means for an aircraft is found in the Type Certificate Data Sheet for that particular model of aircraft. Although the manufacturer's maintenance manual also includes the leveling means for the aircraft, the manual is not FAA approved.

8446. Answer A. JSGT 14-3 (AC 65-9A)
Airworthiness Directives (ADs) are issued by the FAA to notify aircraft owners of an unsafe condition. Compliance with ADs is mandatory, and the conditions for compliance are listed in the AD itself.

8447. Answer A. JSGT 14-4 (Part 43, Appendix A)
Specific criteria to determine whether a job is a major or minor repair is found in FAR Part 43, Appendix A. Major repairs to appliances consist of

1. calibration and repair of instruments.
2. calibration of radio equipment.
3. rewinding the field coil of an electrical accessory.
4. complete disassembly of complex hydraulic power valves.
5. overhaul of pressure type carburetors, and pressure type fuel, oil and hydraulic pumps.

8449. I01
Which maintenance action is an airframe major repair?

A — Changes to the wing or to fixed or movable control surfaces which affect flutter and vibration characteristics.
B — Rewinding the field coil of an electrical accessory.
C — The repair of portions of skin sheets by making additional seams.

8460. I02
An aircraft was not approved for return to service after an annual inspection and the owner wanted to fly the aircraft to another maintenance base. Which statement is correct?

A — The owner must obtain a special flight permit.
B — The aircraft may be flown without restriction up to 10 hours to reach another maintenance base.
C — The owner must obtain a restricted category type certificate.

8461. I02
Each person performing an annual or 100-hour inspection shall use a checklist that contains at least those items in the appendix of

A — 14 CFR Part 43.
B — 14 CFR Part 65.
C — AC 43.13-3

8492. K01
Airworthiness Directives are issued primarily to

A — provide information about malfunction or defect trends.
B — present recommended maintenance procedures for correcting potentially hazardous defects.
C — correct an unsafe condition.

8493. K01
(1) A Supplemental Type Certificate may be issued to more than one applicant for the same design change, providing each applicant shows compliance with the applicable airworthiness requirement.
(2) An installation of an item manufactured in accordance with the Technical Standard Order system requires no further approval for installation in a particular aircraft.
Regarding the above statements,

A — both No. 1 and No. 2 are true.
B — neither No. 1 nor No. 2 is true.
C — only No. 1 is true.

8449. Answer C. JSGT 14-4 (Part 43, Appendix A)
According to FAR Part 43, Appendix A, the repair of portions of skin sheets by making additional seams is an airframe major repair.

8460. Answer A. JSGT 14-5 (FAR 21.197)
If an owner desires to fly an aircraft that is in an unairworthy condition to another maintenance facility where repairs, alterations, or maintenance can be performed, a special flight permit (ferry permit) must be issued.

8461. Answer A. JSGT 14-5 (FAR 43.15)
Each person performing an annual or 100-hour inspection shall use a checklist while performing the inspection. The checklist may be of the person's own design, one provided by the manufacturer of the equipment being inspected, or one obtained from another source. However, the checklist used must include at least the scope and detail of the items contained in Appendix D of FAR Part 43.

8492. Answer C. JSGT 14-3 (AC 65-9A)
Airworthiness Directives (ADs) are used to notify aircraft owners and other interested persons of unsafe conditions as well as methods for correcting the condition.

8493. Answer C. JSGT 14-3 (FAR 21.115)
Only statement number 1 (answer C) is true. Providing the applicant can show compliance with applicable airworthiness requirements, an STC may be issued to any number of applicants for the same design change. Even though an item is manufactured according to a Technical Standard Order, it is not to be interpreted as an approval for installation on any particular aircraft.

Maintenance Publications, Forms, and Records

8494. K01
Primary responsibility for compliance with Airworthiness Directives lies with the

A — Aircraft owner or operator.
B — certificated mechanic holding an Inspection Authorization who conducts appropriate inspections.
C — certificated mechanic who maintains the aircraft.

8495. K01
An aircraft Type Certificate Data Sheet contains

A — maximum fuel grade to be used.
B — control surface adjustment points.
C — location of the datum.

8496. K01
Suitability for use of a specific propeller with a particular engine-airplane combination can be determined by reference to what informational source?

A — Propeller Specifications or Propeller Type Certificate Data Sheet.
B — Aircraft Specifications or Aircraft Type Certificate Data Sheet.
C — Alphabetical Index of Current Propeller Type Certificate Data Sheets, Specifications, and Listings.

8497. K01
When an airworthy (at the time of sale) aircraft is sold, the Airworthiness Certificate

A — becomes invalid until the aircraft is reinspected and returned to service.
B — is voided and a new certificate is issued upon application by the new owner.
C — is transferred with the aircraft.

8498. K01
The issuance of an Airworthiness Certificate is governed by

A — 14 CFR Part 39.
B — 14 CFR Part 21.
C — 14 CFR Part 23.

8499. K01
Specifications pertaining to an aircraft, of which a limited number were manufactured under a type certificate and for which there is no current Aircraft Specifications, can be found in the

A — Aircraft Listing.
B — Annual Summary of Deleted and Discontinued Aircraft Specifications.
C — FAA Statistical Handbook of Civil Aircraft Specifications.

8494. Answer C. JSGT 14-3 (AC 65-9A)
Although a mechanic is responsible for determining that AD's have been complied with during certian airworthiness inspections, the owner or operator of the aircraft is required by CFR Part 91.403 to maintain their aircraft in an airworthy condition, including compliance with applicable AD's.

8495. Answer C. JSGT 14-8 (AC 65-9A)
The only item listed that is on the Type Certificate Data Sheet (TCDS) is the location of the datum. The TCDS also includes, among other things, the minimum fuel grade to be used, and the control surface movements.

8496. Answer B. JSGT 14-8 (AC 65-9A)
The key to this question is matching airplane, engine, and propeller. The only place this information is found is on the Type Certificate Data Sheet for the airplane.

8497. Answer C. JSGT 14-3 (FAR 21.179)
The Airworthiness Certificate is valid as long as the aircraft is maintained in accordance with FAA regulations, and is considered part of the permanent records of the aircraft. When the aircraft is sold, the Airworthiness Certificate is transferred to the new owner.

8498. Answer B. JSGT 14-3 (FAR 21.173)
The issuance of Airworthiness Certificates is covered in FAR Part 21.

8499. Answer A. JSGT 14-12 (AFS 613 Vol VI)
Older aircraft of which fewer than 50 remain in service, or any aircraft of which fewer than 50 were produced or remain in service, are found in the Aircraft Listing.

8500. K01
Where are technical descriptions of certificated propellers found?

A — Applicable Airworthiness Directives.
B — Aircraft Specifications.
C — Propeller Type Certificate Data Sheet.

8501. K01
What information is generally contained in Aircraft Specifications or Type Certificate Data Sheets?

A — Empty weight of the aircraft.
B — Useful load of aircraft.
C — Control surface movements.

8502. K01
Placards required on an aircraft are specified in

A — AC 43.13-1A.
B — FAR's under which the aircraft was type certificated.
C — Aircraft Specifications or Type Certificate Data Sheets.

8503. K01
Technical information about older aircraft models, of which no more than 50 remain in service, can be found in the

A — Aircraft Listing.
B — Annual Summary of Deleted and Discontinued Aircraft Specifications.
C — Alphabetical Index of Antique Aircraft.

8504. K01
(1) The FARs require approval after compliance with the data of a Supplemental Type Certificate.
(2) An installation of an item manufactured in accordance with the Technical Standard Order system requires no further approval for installation in a particular aircraft.
Regarding the above statements,

A — only No. 2 is true.
B — neither No.1 nor No. 2 is true.
C — only No. 1 is true.

8505. K01
Which regulation provides information regarding instrument range markings for an airplane certificated in the normal category?

A — 14 CFR Part 21.
B — 14 CFR Part 25.
C — 14 CFR Part 23.

8500. Answer C. JSGT 14-8 (FAR 21.41)
All specifications for a certificated propeller can be found in the Propeller Type Certificate Data Sheet.

8501. Answer C. JSGT 14-8 (AC 65-9A)
Of the items listed, only the Control Surface Movements are found on the Type Certificate Data Sheet.

8502. Answer C. JSGT 14-8 (AC 65-9A)
The required placards for a particular model aircraft are found in the Aircraft Specifications or Type Certificate Data Sheet for that model.

8503. Answer A. JSGT 14-12 (AFS 613 Vol VI)
Specifications on aircraft of which fewer than 50 were produced, or fewer than 50 remain in service are found in the Aircraft Listing.

8504. Answer C. JSGT 14-12 (Part 21 Subpart E)
Only statement number 1 is correct. Complying with the data of a Supplemental Type Certificate (STC) is a major alteration that must be approved, and a Form 337 completed. Installation of an item manufactured under a Technical Standard Order always requires approval.

8505. Answer C. JSGT 14-3 (FAR 23.1543)
Airworthiness standards for airplanes certificated in the normal, utility, and acrobatic categories are covered in FAR Part 23. One of the standards outlined in this part is instrument markings.

8506.   K01
(1) Propellers are NOT included in the Airworthiness Directive system.
(2) A certificated powerplant mechanic may make a minor repair on an aluminum propeller and approve for return to service.
Regarding the above statements,

A — only No. 2 is true.
B — both No. 1 and No. 2 are true.
C — neither No. 1 nor No. 2 is true.

8507.   K01
An aircraft mechanic is privileged to perform major alterations on U. S. certificated aircraft; however, the work must be done in accordance with FAA-approved technical data before the aircraft can be returned to service. Which is NOT approved data?

A — Airworthiness Directives.
B — AC 43.13-2A.
C — Supplemental Type Certificates.

8508.   K01
What is the maintenance recording responsibility of the person who complies with an Airworthiness Directive?

A — Advise the aircraft owner/operator of the work performed.
B — Make an entry in the maintenance record of that equipment.
C — Advise the FAA district office of the work performed, by submitting an FAA Form 337.

8509.   K01
(1) Manufacturer's data and FAA publications such as Airworthiness Directives, Type Certificate Data Sheets, and advisory circulars are all approved data.
(2) FAA publications such as Technical Standard Orders, Airworthiness Directives, Type Certificate Data Sheets, and Aircraft Specifications and Supplemental Type Certificates are all approved data.
Regarding the above statements,

A — both No. 1 and No. 2 are true.
B — only No. 1 is true.
C — only No. 2 is true.

8506. Answer A. JSGT 14-3 (AC 65-9A) (FAR 65.87)
Only statement number 2 (answer A) is true. A certificated powerplant mechanic with the necessary experience may make minor repairs on aircraft propellers and return them to service. Statement number 1 is incorrect because an Airworthiness Directive may be issued on any product that affects safety. This includes aircraft, engines, propellers, and appliances.

8507. Answer B. JSGT 14-4 (AC 43.9-1E)
Airworthiness Directives, Type Certificates, and FAA-approved data provided by the manufacturer are all approved data. AC 43.13-2A (answer B) provides acceptable methods of performing aircraft alterations but is not considered approved data. A good practice when planning an alteration that is not specifically covered by an AD, STC, or other approved data is to outline your work and submit a pencil copy of a 337 Form to your local FAA maintenance inspector for approval.

8508. Answer B. JSGT 14-4 (AC 65-9A)
Compliance with Airworthiness Directives is required, and a record of the action taken to comply with an AD must be included in the permanent records of the equipment affected by the AD.

8509. Answer C. JSGT 14-3 (AC 43.9-1E)
Only statement number 2 (answer C) is true. Items such as Technical Standard Orders, Airworthiness Directives, Type Certificate Data Sheets, and Aircraft Specifications and Supplemental Type Certificates are all approved data. Items such as advisory circulars are not considered approved data.

8510.   **K01**
The Air Transport Association of America (ATA) Specification No. 100
(1) establishes a standard for the presentation of technical data in maintenance manuals.
(2) divides the aircraft into numbered systems and subsystems in order to simplify locating maintenance instructions.
Regarding the above statements,

A — both No. 1 and No. 2 are true.
B — neither No. 1 nor No. 2 is true.
C — only No. 1 is true.

8510. Answer A. JSGT 14-12 (AC 65-9A)
Both statements are correct. The ATA Specification 100 is a standard for the presentation of technical information. Because of this specification, maintenance information from any of the manufacturers of transport aircraft is arranged in the same way.

8511.   **K01**
General Aviation Airworthiness Alerts

A — provide mandatory procedures to prevent or correct serious aircraft problems.
B — provide information about aircraft problems and suggested corrective actions.
C — provide temporary emergency procedures until Airworthiness Directives can be issued.

8511. Answer B. JSGT 14-8 (AC 43.16)
The General Aviation Alert system (AC 43.16) is published monthly to inform maintenance personnel about aircraft problems and suggested corrective actions.

8515.   **K02**
(Refer to figure 63) An aircraft has a total time in service of 468 hours. The Airworthiness Directive given was initially complied with at 454 hours in service. How many additional hours in service may be accumulated before the Airworthiness Directive must again be complied with?

A — 46.
B — 200.
C — 186.

8515. Answer C. JSGT 14-7 (AC 65-9A)
Although compliance wasn't required until 500-hours total time, the AD was complied with at 454-hours total time. Once the AD is complied with, you are required to abide by the repetitive 200-hour inspection. The aircraft has been in service for 14 hours (468 − 454 = 14) since the AD was complied with, and, therefore, the AD must again be complied with in 186 hours (200 − 14 = 186).

The following is the compliance portion of an Airworthiness Directive. "Compliance required as indicated, unless already accomplished:

I. Aircraft with less than 500-hours' total time in service: Inspect in accordance with instructions below 500-hours' total time, or within the next 50-hours' time in service after the effective date of this AD, and repeat after each subsequent 200 hours in service.

II. Aircraft with 500-hours' through 1,000-hours' total time in service: Inspect in accordance with instructions below within the next 50-hours' time in service after the effective date of this AD, and repeat after each subsequent 200 hours in service.

III. Aircraft with more than 1,000-hours' time in service: Inspect in accordance with instructions below within the next 25-hours' time in service after the effective date of this AD, and repeat after each subsequent 200 hours in service."

FIGURE 63.—Airworthiness Directive Excerpt.

8516. K02

The following is a table of airspeed limits as given in an FAA-issued aircraft specification.

Normal operating speed ....260 knots
Never-exceed speed...293 knots
Maximum landing gear
operating speed ....174 knots
Maximum flap extended speed ....139 knots

The high end of the white arc on the airspeed instrument would be at

A — 260 knots.
B — 174 knots.
C — 139 knots.

8516. Answer C. JSGT 14-3 (FAR 23.1545)

According to FAR Part 23, the flap operating speed range on an airspeed indicator is marked with a white arc. The high end of this arc designates the maximum flap extension speed.

8517. K02

A complete detailed inspection and adjustment of the valve mechanism will be made at the first 25 hours after the engine has been placed in service. Subsequent inspections of the valve mechanism will be made each second 50-hour period.
From the above statement, at what intervals will valve mechanism inspections be performed?

A — 100 hours.
B — 50 hours.
C — 125 hours.

8517. Answer A. JSGT 14-7 (AC 65-19F)

The statement indicates that after the first inspection, subsequent inspections are to be made at each SECOND 50-hour period. Two 50-hour periods equals 100 hours.

8518. K02

Check thrust bearing nuts for tightness on new or newly overhauled engines at the first 50-hour inspection following installation. Subsequent inspections on thrust bearing nuts will be made at each third 50-hour inspection.
From the above statement, at what intervals should you check the thrust bearing nut for tightness?

A — 150 hours.
B — 200 hours.
C — 100 hours.

8518. Answer A. JSGT 14-7 (AC 65-19F)

The statement identifies the interval for subsequent inspections as each third 50-hour period. Three 50-hour periods is equivalent to 150 hours.

8520. L01

A repair, as performed on an airframe, shall mean

A — the upkeep and preservation of the airframe including the component parts thereof.
B — the restoration of the airframe to a condition for safe operation after damage or deterioration.
C — simple or minor preservation operations and the replacement of small standard parts not involving complex assembly operations.

8520. Answer B. JSGT 14-4 (Part 43, Appendix A)

Restoration after damage or deterioration is a repair (answer B). The upkeep and preservation, as well as simple replacement of small parts are preventive maintenance (answers A and C).

8521. L01

The replacement of fabric on fabric-covered parts such as wings, fuselages, stabilizers, or control surfaces is considered to be a

A — minor repair unless the new cover is different in any way from the original cover.
B — minor repair unless the underlying structure is altered or repaired.
C — major repair even though no other alteration or repair is performed.

8522. L01

Which is classified as a major repair?

A — The splicing of skin sheets.
B — Installation of new engine mounts obtained from the aircraft manufacturer.
C — Any repair of damaged stressed metal skin.

8527. L01

The replacement of a damaged vertical stabilizer with a new identical stabilizer purchased from the aircraft manufacturer is considered a

A — minor alteration.
B — major repair.
C — minor repair.

8533. L01

Who is responsible for determining that materials used in aircraft maintenance and repair are of the proper type and conform to the appropriate standards?

A — The installing person or agency.
B — The owner of the aircraft.
C — The manufacturer of the aircraft.

8534. L01

Which of these publications contains standards for protrusion of bolts, studs, and screws through self-locking nuts?

A — AC 43.13-2.
B — Aircraft Specifications or Type Certificate Data Sheets.
C — AC 43.13-1B.

8535. L01

The replacement of a damaged engine mount with a new identical engine mount purchased from the aircraft manufacturer is considered a

A — minor alteration.
B — major repair.
C — minor repair.

8521. Answer C. JSGT 14-4 (Part 43, Appendix A)
It is clearly stated in FAR Part 43, Appendix A that replacement of fabric covering is a major repair.

8522. Answer A. JSGT 14-4 (Part 43, Appendix A)
Making additional seams or splicing skin sheets to repair a portion of the skin is a major repair. Answers (B) and (C) are incorrect because they are both considered minor repairs.

8527. Answer C. JSGT 14-5 (Part 43, Appendix A)
Generally, the replacement of a complete assembly with an identical assembly purchased from the manufacturer is a minor repair, regardless of the size of the part. One way to determine if something constitutes a major or minor repair is that if the action is not included in FAR Part 43, Appendix A, it is a minor repair or alteration.

8533. Answer A. JSGT 14-3 (FAR 43.13)
Whenever maintenance is being performed, it is the responsibility of the technician performing the work to ensure that the parts being used meet appropriate standards.

8534. Answer C. JSGT 14-7 (AC 43.13-1A)
AC 43.13-1B provides a wealth of information on standard applications for hardware, as well as torque charts and other information.

8535. Answer C. JSGT 14-5 (Part 43, Appendix A)
Generally, if you are replacing a part with an identical new part, the operation is a minor repair. This could change if the part required welding or riveting for installation.

Maintenance Publications, Forms, and Records   14-9

SECTION B
FORMS AND RECORDS

Section B of Chapter 14 discusses the steps required to properly document aircraft maintenance. Included is information on inspection checklists, aircraft logbooks, and common phraseology used to sign off an aircraft and return it to service. The following FAA Test questions are taken from this section:

8443, 8444, 8445, 8448, 8450, 8453, 8454, 8456, 8457, 8458, 8459, 8462, 8463.

8443.   I01
Where is the record of compliance with Airworthiness Directives or manufacturers' service bulletins normally indicated?

A — FAA Form 337.
B — Aircraft maintenance records.
C — Flight manual.

8443. Answer B. JSGT 14-21 (AC 65-9A)
Record of compliance with Airworthiness Directives and manufacturer's service bulletins must be entered into the aircraft's maintenance records.

8444.   I01
If work performed on an aircraft has been done satisfactorily, the signature of an authorized person on the maintenance records for maintenance or alterations performed constitutes

A — approval of the aircraft for return to service.
B — approval for return to service only for the work performed.
C — only verification that the maintenance or alterations were performed referencing maintenance data.

8444. Answer B. JSGT 14-22 (FAR 43.9)
The signature of the person approving the work constitutes approval for return to service only for the work performed.

8445.   I01
During an annual inspection, if a defect is found which makes the aircraft unairworthy, the person disapproving must

A — void the aircraft's Airworthiness Certificate.
B — submit a Malfunction or Defect Report.
C — provide a written notice of the defect to the owner.

8445. Answer C. JSGT 14-23 (FAR 43.11)
If a defect is found during an inspection that renders the aircraft unairworthy, regulations require the person disapproving the inspection to provide a signed and dated list of discrepancies to the owner.

8448.   I01
Where should you find this entry?
"Removed right wing from aircraft and removed skin from outer 6 feet. Repaired buckled spar 49 inches from tip in accordance with figure 8 in the manufacturer's structural repair manual No. 28-1"

A — Aircraft engine maintenance record.
B — Aircraft minor repair and alteration record.
C — FAA Form 337

8448. Answer C. JSGT 14-16 (Part 43, Appendix A and B)
This entry is describing a major repair based on FAR Part 43, Appendix A criteria. All major repairs must be recorded on an FAA Form 337.

8450.   I01
Which aircraft record entry is the best description of the replacement of several damaged heli-coils in a casting?

A — Eight 1/4 - 20 inch standard heli-coils were replaced. The damaged inserts were extracted, the tapped holes gauged, then new inserts installed, and tangs removed.
B — Eight 1/4 - 20 inch standard heli-coils were installed in place of damaged ones.
C — Eight 1/4 - 20 inch standard heli-coil inserts were repaired by replacing the damaged inserts with a lock-type insert, after the tapped holes were checked for corrosion.

8450. Answer A. JSGT 14-22 (AC 65-9A)
This question is somewhat subjective. However, when making logbook entries you should describe the work accomplished so that the reader will know, without a doubt, what you did, and what parts you used in the process. Answer (A) provides the most complete description of the operations involved in replacing a standard heli-coil.

8453.   I01
Which aircraft record entry best describes a repair of a dent in a tubular steel structure dented at a cluster?

A — Removed and replaced the damaged member.
B — Welded a reinforcing plate over the dented area.
C — Filled the damaged area with a molten metal and dressed to the original contour.

8453. Answer B. JSGT 14-22 (AC 43.13-1A)
To repair dents at a steel-tube cluster-joint, weld a specially formed steel patch plate over the dented area and surrounding tubes.

8454.   I02
Who is responsible for making the entry in the maintenance records after an annual, 100 hour, or progressive inspection?

A — The owner or operator of the aircraft.
B — The person approving or disapproving for return to service.
C — The designee or inspector representing the FAA Administrator.

8454. Answer B. JSGT 14-22 (FAR 43.11)
Regardless of how many mechanics there are working on an aircraft, the person responsible for the logbook entry is the person approving or disapproving the aircraft for return to service.

8456.   I02
When approving for return to service after maintenance or alteration, the approving person must enter in the maintenance record of the aircraft

A — the date the maintenance or alteration was begun, a description (or reference to acceptable data) of work performed, the name of the person performing the work (if someone else), signature, and certificate number.
B — a description (or reference to acceptable data) of work performed, date of completion, the name of the person performing the work (if someone else), signature, and certificate number.
C — a description (or reference to acceptable data) of work performed, date of completion, the name of the person performing the work (if someone else), signature, certificate number, and kind of certificate held.

8456. Answer C. JSGT 14-22 (FAR 43.9)
Each person who maintains, performs preventative maintenance, rebuilds, or alters an aircraft airframe, aircraft engine, propeller, appliance, or component shall make an entry in the maintenance record containing the following information.

1. A description (or reference to acceptable data) of work performed.
2. Date the work was completed.
3. The name of the person performing the work (if someone else).
4. Signature.
5. Certificate number.
6. Kind of certificate held.

8457. I02
What action is required when a minor repair is performed on a certificated aircraft?

A — An FAA Form 337 must be completed.
B — An entry in the aircraft's maintenance record is required.
C — The owner of the aircraft must annually report minor repairs to the FAA.

8458. I02
After making a certain repair to an aircraft engine that is to be returned to service, an FAA Form 337 is prepared. How many copies are required and what is the disposition of the completed forms?

A — Two; one copy for the aircraft owner and one copy for the FAA.
B — Two; one copy for the FAA and one copy for the permanent records of the repairing agency or individual.
C — Three; one copy for the aircraft owner, one copy for the FAA, and one copy for the permanent records of the repairing agency or individual.

8459. I02
Who is responsible for upkeep of the required maintenance records for an aircraft?

A — The maintaining repair station or authorized inspector.
B — The maintaining certificated mechanic.
C — The aircraft owner.

8462. I02
An FAA Form 337 is used to record and document

A — preventive and routine maintenance.
B — major and minor repairs, and major and minor alterations.
C — major repairs and major alterations.

8463. I02
After a mechanic holding an airframe and powerplant rating completes a 100-hour inspection, what action is required before the aircraft is returned to service?

A — Make the proper entries in the aircraft's maintenance record.
B — An operational check of all systems.
C — A mechanic with an inspection authorization must approve the inspection.

8457. Answer B. JSGT 14-22 (FAR 43.9; Part 43, Appendix B)
All major repairs and major alterations must be accompanied by an FAA Form 337 and entered in the aircraft's permanent records. However, minor repairs and minor alterations only need to be entered in the aircraft's permanent records.

8458. Answer A. JSGT 14-16 (Part 43, Appendix B)
The key here is that the question asks for the required number of copies. Regulations require that you give a signed copy to the aircraft owner and forward a copy to the local Flight Standards District Office (FSDO) within 48 hours after the part is approved for return to service.

8459. Answer C. JSGT 14-21 (FAR 91.417)
The person ultimately responsible for maintaining an aircraft in an airworthy condition, including its maintenance records, is the aircraft owner.

8462. Answer C. JSGT 14-16 (Part 43, Appendix B)
FAA Form 337 is used to document any major repair or major alteration to an aircraft, engine, propeller, or appliance

8463. Answer A. JSGT 14-22 (FAR 43.11)
Once a 100-hour inspection is complete, the mechanic must make the appropriate entries in the aircraft's maintenance records before the aircraft can be returned to service.

MECHANIC PRIVILEGES AND LIMITATIONS

SECTION A
THE MECHANIC CERTIFICATE

Chapter 15 discusses the privileges and limitations associated with Mechanic's Certificates, Repairman's Certificates, and the Inspection Authorization. FAA Test questions based on this section include:

8455, 8464, 8519, 8523, 8524, 8525, 8526, 8528, 8529, 8530, 8531, 8532, 8536, 8537.

8455. I02
An aircraft owner was provided a list of discrepancies on an aircraft that was not approved for return to service after an annual inspection. Which statement is correct concerning who may correct the discrepancies?

A — Only a mechanic with an inspection authorization.
B — An appropriately rated mechanic.
C — Any certificated repair station.

8464. I02
Which of the following may a certificated airframe and powerplant mechanic perform on aircraft and approve for return to service?

1. a 100-hour inspection.
2. an annual inspection.
3. a progressive inspection.

A.— 1, 3.
B.— 1, 2.
C.— 1, 2, 3.

8455. Answer B. JSGT 15-3 (FAR 43.3)
An A&P technician must hold an inspection authorization to perform an annual inspection. However, discrepancies may be corrected by any mechanic who is appropriately rated for the work required to correct the discrepancy.

8464. Answer A. JSGT 15-3 (FAR 65.85, 65.87)
Provided an airframe and powerplant mechanic is current and has the necessary experience, they may return an aircraft to service following a 100-hour inspection. For an annual inspection (choice 2) and a progressive inspection (choice 3), the mechanic must also have an Inspection Authorization.

8519. **L01**
Certificated mechanics with a powerplant rating may perform the

A — annual inspection required by the Federal Aviation Regulations on a powerplant or propeller or any component thereof, and may release the same to service.
B — 100-hour and/or annual inspections required by the Federal Aviation Regulations on powerplants, propellers, or any components thereof, and may release the same to service.
C — 100-hour inspection required by the Federal Aviation Regulations on a powerplant, propeller, or any component thereof, and may release the same to service.

8519. Answer C. JSGT 15-2 (FAR 65.87)
A certificated mechanic with a powerplant rating may approve and return to service a powerplant or propeller or any related part or appliance, after performing a 100-hour inspection required by Part 91. Both answers (A) and (B) are incorrect because an annual inspection must be performed by a mechanic with an Inspection Authorization.

8523. **L01**
The 100-hour inspection required by FAR's for certain aircraft being operated for hire may be performed by

A — persons working under the supervision of an appropriately rated mechanic, but the aircraft must be approved by the mechanic for return to service.
B — appropriately rated mechanics only if they have an inspection authorization.
C — appropriately rated mechanics and approved by them for return to service.

8523. Answer C. JSGT 15-2 (FAR 65.85 and 65.87)
According to FAR Part 65, aircraft mechanics with Airframe and Powerplant ratings may perform a 100-hour inspection and return an aircraft to service. Persons working under the supervision of an appropriately rated mechanic (answer A) may not, according to FAR 43.3, conduct a 100-hour inspection. Answer (B) is incorrect because you do not need an inspection authorization to conduct a 100-hour inspection.

8524. **L01**
A person working under the supervision of a certificated mechanic with an airframe and powerplant rating is not authorized to perform

A — repair of a wing brace strut by welding.
B — a 100-hour inspection.
C — repair of an engine mount by riveting.

8524. Answer B. JSGT 15-3 (FAR 43.3)
A person working under the supervision of a certificated mechanic may perform the maintenance, preventive maintenance, and alterations that his supervisor is authorized to perform. However, this authorization is not extended to any inspections required by Part 91 or Part 125. This includes 100-hour inspections.

8525. **L01**
Certificated mechanics, under their general certificate privileges, may

A — perform minor repairs to instruments.
B — perform 100-hour inspection of instruments.
C — perform minor alterations to instruments.

8525. Answer B. JSGT 15-3 (Part 43, Appendix D; FAR 65.81)
Instrument repairs and alterations must be accomplished by an appropriately rated repair station. The mechanic can only inspect the instruments within the scope of the 100-hour inspection. This includes inspecting for poor condition, mounting, marking, and where practicable, for improper operation.

Mechanic Privileges and Limitations 15-3

8526. L01
An Airworthiness Directive requires that a propeller be altered. Certificated mechanics could

A — perform and approve the work for return to service if it is a minor alteration.
B — not perform the work because it is an alteration.
C — not perform the work because they are not allowed to perform and approve for return to service, repairs or alterations to propellers.

8528. L01
FAA certificated mechanics may

A — approve for return to service a major repair for which they are rated.
B — supervise and approve a 100-hour inspection.
C — approve for return to service a minor alteration they have performed appropriate to the rating(s) they hold.

8529. L01
A certificated mechanic with a powerplant rating may perform the

A — annual inspection required by the FAR's on a powerplant or any component thereof and approve and return the same to service.
B — 100-hour inspection required by the FAR's on a powerplant or any component thereof and approve and return the same to service.
C — 100-hour inspection required by the FAR's on an airframe, powerplant, or any component thereof and approve and return the same to service.

8530. L01
What part of the FAR's prescribes the requirements for issuing mechanic certificates and associated ratings and the general operating rules for the holders of these certificates and ratings?

A — 14 CFR Part 43.
B — 14 CFR Part 91.
C — 14 CFR Part 65.

8526. Answer A. JSGT 15-3 (FAR 65.81 and 65.87)
Powerplant certificated mechanics may perform minor repairs and minor alterations on propellers and return them to service.

8528. Answer C. JSGT 15-3 (FAR 65.81)
FAA-certified machanics are authorized to perform minor alterations appropriate to the rating(s) they hold, and return them to service

8529. Answer B. JSGT 15-3 (FAR 65.87)
FAR Part 65 authorizes the powerplant certificated mechanic to perform a 100-hour inspection on a powerplant, propeller, and component parts thereof, and approve the items for return to service.

8530. Answer C. JSGT 15-2 (FAR 65.1)
FAR Part 65 covers the certification requirements for all airmen other than flight crewmembers. This includes mechanics and repairmen.

8531. L01
A certificated mechanic shall not exercise the privileges of the certificate and rating unless, within the preceding 24 months, the Administrator has found that the certificate holder is able to do the work or the certificate holder has

A — served as a mechanic under the certificate and rating for at least 18 months.
B — served as a mechanic under the certificate and rating for at least 12 months.
C — served as a mechanic under the certificate and rating for at least 6 months.

8531. Answer C. JSGT 15-3 (FAR 65.83)
To be considered current, a certificated mechanic must, for at least 6 months out of 24, serve as a mechanic under his certificate and rating.

8532. L01
(1) Certificated mechanics with an airframe rating may perform a minor repair to an airspeed indicator providing they have the necessary equipment available.
(2) Certificated mechanics with a powerplant rating may perform a major repair to a propeller providing they have the necessary equipment available.
Regarding the above statements,

A — only No. 1 is true.
B — neither No. 1 nor No. 2 is true.
C — only No. 2 is true.

8532. Answer B. JSGT 15-3 (FAR 65.81)
Neither of these statements is true (answer B). Even minor repairs to instruments must be accomplished by a certified instrument repair station and certificated powerplant mechanics may only perform MINOR repairs to propellers.

8536. L01
Who has the authority to approve for return to service a powerplant or propeller or any part thereof after a 100-hour inspection?

A — A mechanic with a powerplant rating.
B — Any certificated repairman.
C — Personnel of any certificated repair station.

8536. Answer A. JSGT 15-3 (FAR 65.87)
A certificated mechanic with the powerplant rating is authorized to perform a 100-hour inspection and return to service a powerplant, propeller, or any part thereof.

8537. L01
Instrument repairs may be performed

A — by the instrument manufacturer only.
B — by an FAA-approved instrument repair station.
C — on airframe instruments by mechanics with an airframe rating.

8537. Answer B. JSGT 15-4 (AC 43.13-1A)
Repair and overhaul of aircraft instruments must be made by an FAA-approved facility having proper test equipment, adequate manufacturer's maintenance manuals and service bulletins, and qualified personnel.

SUBJECT MATTER KNOWLEDGE CODES

APPENDIX 1

To determine the knowledge area in which a particular question was incorrectly answered, compare the subject matter code(s) on AC Form 8080-2, Airmen Written Test Report, to the subject matter outline that follows. The total number of test items missed may differ from the number of subject matter codes shown on the AC Form 8080-2, since you may have missed more than one question in a certain subject matter code.

Basic Electricity

A01	Calculate and measure capacitance and inductance
A02	Calculate and measure electrical power
A03	Measure voltage, current, resistance, and continuity
A04	Determine the relationship of voltage, current, and resistance in electrical circuits
A05	Read and interpret electrical circuit diagrams, including solid state devices and logic functions
A06	Inspect and service batteries

Aircraft Drawings

B01	Use drawings, symbols, and system schematics
B02	Draw sketches of repairs and alterations
B03	Use blueprint information
B04	Use graphs and charts

Weight and Balance

C01	Weigh aircraft
C02	Perform complete weight-and-balance check and record data

Fluid Lines and Fittings

D01	Fabricate and install rigid and flexible fluid lines and fittings

Materials and Processes

E01	Identify and select appropriate nondestructive testing methods
E02	Perform dye penetrant, eddy current, ultrasonic, and magnetic particle inspections
E03	Perform basic heat-treating processes
E04	Identify and select aircraft hardware and materials
E05	Inspect and check welds
E06	Perform precision measurements

Ground Operation and Servicing

F01	Start, ground operate, move, service, and secure aircraft and identify typical ground operation hazards
F02	Identify and select fuels

Cleaning and Corrosion Control

G01	Identify and select cleaning materials
G02	Inspect, identify, remove, and treat aircraft corrosion and perform aircraft cleaning

Mathematics

H01	Extract roots and raise numbers to a given power
H02	Determine areas and volumes of various geometrical shapes
H03	Solve ratio, proportion, and percentage problems
H04	Perform algebraic operations involving addition, subtraction, multiplication, and division of positive and negative numbers

Maintenance Forms and Records

I01   Write descriptions of work performed including aircraft discrepancies and corrective actions using typical aircraft maintenance records

I02   Complete required maintenance forms, records, and inspection reports

Basic Physics

J01   Use and understand the principles of simple machines; sound, fluid, and heat dynamics; basic aerodynamics; aircraft structures; and theory of flight

Maintenance Publications

K01   Demonstrate ability to read, comprehend, and apply information contained in FAA and manufacturer's aircraft maintenance specifications, data sheets, manuals, publications, and related Federal Aviation Regulations, Airworthiness Directives, and Advisory material

K02   Read technical data

Mechanic Privileges and Limitations

L01   Exercise mechanic privileges within the limitations prescribed by FAR Part 65

NOTE: AC 00-2, Advisory Circular Checklist, transmits the status of all FAA advisory circulars (ACs), as well as FAA internal publications and miscellaneous flight information such as AIM, Airport/Facility Directory, written test question books, and other material directly related to a certificate or rating. To obtain a free copy of the AC 00-2, send your request to:

U.S. Department of Transportation
Utilization and Storage Section, M-443.2
Washington, DC 20590

CROSS-REFERENCE LISTING OF QUESTIONS

APPENDIX 2

Appendix 2 is a numerical listing of all the Aviation Mechanic General Knowledge questions found in the *Aviation Mechanic General Written Test Book*, FAA-T-8080-10E. The listing includes the FAA question number and the answer. The cross-reference listing is to the right of the answer. It refers to the chapter and the page in the *A&P Technician General Textbook* where the question is answered.

Example: 8478 C 2-5

This indicates that the answer to question 8478 is C, and the question is answered in Chapter 2, page 2-5 of the Study Guide.

QUESTION	ANSWER	PAGE	QUESTION	ANSWER	PAGE	QUESTION	ANSWER	PAGE
8001	C	3-19	8035	A	3-10	8069	C	3-32
8002	C	3-20	8036	C	3-10	8070	A	3-33
8003	A	3-20	8037	B	3-3	8071	A	3-33
8004	C	3-20	8038	B	3-11	8072	C	3-33
8005	A	3-20	8039	B	3-11	8073	C	3-34
8006	A	3-20	8040	B	3-24	8074	A	3-6
8007	C	3-20	8041	A	3-23	8075	A	3-24
8008	C	3-21	8042	C	3-11	8076	C	3-24
8009	C	3-21	8043	A	3-11	8077	C	3-24
8100	B	3-21	8044	C	3-12	8078	A	3-25
8011	C	3-22	8045	C	3-12	8079	B	3-26
8012	C	3-22	8046	C	3-13	8080	A	3-26
8013	A	3-22	8047	B	3-13	8081	A	3-26
8014	A	3-22	8048	A	3-23	8082	B	3-26
8015	C	3-7	8049	B	3-14	8083	C	3-27
8016	C	3-1	8050	C	3-14	8084	B	3-28
8017	A	3-1	8051	B	3-4	8085	A	3-16
8018	C	3-7	8052	B	3-4	8086	C	3-16
8019	C	3-2	8053	A	3-15	8087	A	3-16
8020	A	3-2	8054	A	3-15	8088	B	3-16
8021	C	3-7	8055	C	3-4	8089	B	3-17
8022	C	3-23	8056	B	3-4	8090	C	3-17
8023	C	3-2	8057	A	3-29	8091	C	3-17
8024	C	3-23	8058	A	3-29	8092	B	3-17
8025	B	3-8	8059	B	3-29	8093	C	3-17
8026	A	3-8	8060	C	3-30	8094	A	3-18
8027	C	3-9	8061	B	3-30	8095	C	3-18
8028	C	3-2	8062	A	3-31	8096	B	3-18
8029	C	3-3	8063	A	3-31	8097	C	3-18
8030	C	3-3	8064	C	3-31	8098	C	3-18
8031	A	3-9	8065	B	3-4	8099	B	3-18
8032	B	3-9	8066	C	3-4	8100	B	3-19
8033	B	3-3	8067	A	3-32	8101	B	3-19
8034	B	3-10	8068	B	3-32	8102	A	3-19

QUESTION	ANSWER	PAGE	QUESTION	ANSWER	PAGE	QUESTION	ANSWER	PAGE
8103	A	5-9	8158	A	6-2	8213	C	10-4
8104	B	5-10	8159	C	14-1	8214	B	10-4
8105	C	5-1	8160	A	6-2	8215	B	10-4
8106	B	5-2	8161	A	6-2	8216	B	2-3
8107	C	5-10	8162	A	6-2	8217	B	10-5
8108	C	5-2	8163	C	6-3	8218	B	10-6
8109	C	5-2	8164	C	6-9	8219	C	11-1
8110	B	5-10	8165	C	6-3	8220	C	11-1
8111	A	5-3	8166	B	6-3	8221	C	11-7
8112	C	5-4	8167	C	6-3	8222	A	11-8
8113	A	5-11	8168	B	6-3	8223	C	11-1
8114	B	5-11	8169	C	6-3	8224	B	11-8
8115	B	5-12	8170	A	6-4	8225	A	11-2
8116	B	5-4	8171	A	6-4	8226	C	11-2
8117	C	5-12	8172	A	6-9	8227	B	11-8
8118	B	5-13	8173	B	6-4	8228	B	11-2
8119	B	5-4	8174	B	6-6	8229	C	11-2
8120	A	5-13	8175	C	6-9	8230	B	11-2
8121	B	5-13	8176	B	6-4	8231	A	11-2
8122	C	5-13	8177	C	6-7	8232	A	11-3
8123	C	5-14	8178	C	6-4	8233	B	11-3
8124	A	5-14	8179	B	6-5	8234	A	11-3
8125	B	5-15	8180	B	6-7	8235	A	11-3
8126	A	5-15	8181	A	6-7	8236	C	11-3
8127	C	5-15	8182	B	6-5	8237	A	11-4
8128	A	5-15	8183	A	6-5	8238	C	11-4
8129	C	5-16	8184	B	6-5	8239	A	11-4
8130	A	5-16	8185	C	6-8	8240	C	11-4
8131	A	5-4	8186	A	6-6	8241	B	11-4
8132	B	5-16	8187	C	6-8	8242	A	11-5
8133	A	5-17	8188	C	6-8	8243	A	11-5
8134	B	5-4	8189	B	6-8	8244	C	11-5
8135	C	5-17	8190	B	6-9	8245	B	12-3
8136	B	5-4	8191	C	6-6	8246	A	7-1
8137	A	5-5	8192	C	10-1	8247	A	7-1
8138	C	5-5	8193	A	10-1	8248	C	7-1
8139	C	5-5	8194	B	10-1	8249	B	7-1
8140	B	5-5	8195	A	10-2	8250	C	7-2
8141	B	5-17	8196	C	10-2	8251	B	7-2
8142	A	5-18	8197	C	10-5	8252	C	7-2
8143	B	5-19	8198	B	10-2	8253	A	7-2
8144	A	5-19	8199	A	10-2	8254	C	7-2
8145	A	5-20	8200	C	10-2	8255	B	7-2
8146	A	5-20	8201	C	10-5	8256	A	8-1
8147	A	5-21	8202	C	10-6	8257	A	7-3
8148	B	5-21	8203	C	10-3	8258	A	8-1
8149	A	5-21	8204	A	10-3	8259	B	7-3
8150	B	5-22	8205	B	10-3	8260	C	8-1
8151	A	5-22	8206	A	10-3	8261	C	7-3
8152	C	5-23	8207	A	10-3	8262	C	8-2
8153	A	6-1	8208	C	10-4	8263	C	8-2
8154	C	6-1	8209	C	10-6	8264	B	8-2
8155	A	6-1	8210	A	10-4	8265	B	8-2
8156	A	6-2	8211	B	10-6	8266	C	8-3
8157	A	6-2	8212	B	10-4	8267	B	8-3

Cross-Reference Listing of Questions

QUESTION	ANSWER	PAGE	QUESTION	ANSWER	PAGE	QUESTION	ANSWER	PAGE
8268	C	8-3	8323	C	13-5	8378	C	12-5
8269	A	8-3	8324	B	13-6	8379	C	1-1
8270	A	8-4	8325	A	13-7	8380	A	1-1
8271	C	8-4	8326	C	13-5	8381	C	1-8
8272	B	8-4	8327	A	13-5	8382	C	1-1
8273	A	7-3	8328	B	13-5	8383	C	1-1
8274	C	7-3	8329	B	13-5	8384	C	1-2
8275	C	8-4	8330	A	13-5	8385	B	1-2
8276	A	7-3	8331	C	13-6	8386	A	1-2
8277	A	8-4	8332	B	13-6	8387	C	1-2
8278	C	11-5	8333	C	13-6	8388	A	1-2
8279	B	11-6	8334	A	13-6	8389	C	1-2
8280	A	7-4	8335	A	13-7	8390	C	1-2
8281	B	11-6	8336	C	13-7	8391	C	1-3
8282	A	11-6	8337	B	13-7	8392	A	1-3
8283	C	11-6	8338	C	13-7	8393	C	1-3
8284	A	11-6	8339	A	13-7	8394	B	1-11
8285	B	11-6	8340	B	13-8	8395	A	1-11
8286	B	11-6	8341	C	13-8	8396	C	1-11
8287	A	11-7	8342	C	13-8	8397	B	1-12
8288	B	11-7	8343	C	13-8	8398	A	2-3
8289	A	9-1	8344	A	13-8	8399	A	1-12
8290	A	9-2	8345	B	13-8	8400	B	1-12
8291	C	9-2	8346	C	13-8	8401	C	1-13
8292	B	9-3	8347	A	12-1	8402	B	1-13
8293	B	9-3	8348	C	12-1	8403	B	1-13
8294	C	9-3	8349	B	12-5	8404	B	1-13
8295	A	9-4	8350	A	12-1	8405	B	1-14
8296	B	9-4	8351	A	3-19	8406	A	1-14
8297	C	9-5	8352	B	12-1	8407	C	1-14
8298	C	9-5	8353	A	12-2	8408	B	1-14
8299	A	9-5	8354	B	12-2	8409	A	1-3
8300	A	9-5	8355	C	12-2	8410	C	1-4
8301	B	9-6	8356	A	12-3	8411	C	1-4
8302	B	9-6	8357	A	12-3	8412	C	1-4
8303	C	9-6	8358	C	12-5	8413	C	1-4
8304	C	9-6	8359	C	12-3	8414	A	1-4
8305	B	9-6	8360	C	12-3	8415	A	1-5
8306	B	9-6	8361	B	12-6	8416	A	1-5
8307	B	9-6	8362	A	7-4	8417	C	1-5
8308	A	13-1	8363	C	12-4	8418	C	1-5
8309	B	13-1	8364	B	12-6	8419	B	1-5
8310	C	13-1	8365	A	12-4	8420	C	1-5
8311	C	13-2	8366	C	12-6	8421	B	1-6
8312	C	13-2	8367	B	12-6	8422	C	1-6
8313	C	13-2	8368	B	12-2	8423	B	1-6
8314	B	13-2	8369	B	12-2	8424	C	1-6
8315	A	13-3	8370	A	12-6	8425	A	1-6
8316	C	13-3	8371	B	12-4	8426	B	1-6
8317	A	13-3	8372	C	12-4	8427	A	1-7
8318	A	13-3	8373	A	12-6	8428	B	1-7
8319	C	13-4	8374	C	12-4	8429	A	1-7
8320	B	13-4	8375	C	12-4	8430	A	1-7
8321	B	13-4	8376	B	12-4	8431	A	3-6
8322	B	13-4	8377	B	12-5	8432	B	1-8

QUESTION	ANSWER	PAGE	QUESTION	ANSWER	PAGE	QUESTION	ANSWER	PAGE
8433	C	1-8	8468	A	2-1	8503	A	14-4
8434	C	1-9	8469	C	2-5	8504	C	14-4
8435	A	1-7	8470	C	2-4	8505	C	14-4
8436	A	1-9	8471	B	2-2	8506	A	14-5
8437	C	1-9	8472	B	2-5	8507	B	14-5
8438	B	1-7	8473	B	2-5	8508	B	14-5
8439	A	1-10	8474	B	2-7	8509	C	14-5
8440	C	1-10	8475	C	2-4	8510	A	14-6
8441	A	1-10	8476	C	2-4	8511	B	14-6
8442	A	1-10	8477	B	2-5	8512	B	5-5
8443	B	14-9	8478	C	2-5	8513	A	5-5
8444	B	14-9	8479	B	2-4	8514	A	5-5
8445	C	14-9	8480	C	2-2	8515	C	14-6
8446	A	14-1	8481	A	2-3	8516	C	14-7
8447	A	14-1	8482	B	2-4	8517	A	14-7
8448	C	14-9	8483	B	2-6	8518	A	14-7
8449	C	14-2	8484	A	2-6	8519	C	15-2
8450	A	14-10	8485	A	2-5	8520	B	14-7
8451	A	8-4	8486	C	2-6	8521	C	14-8
8452	C	10-5	8487	C	2-6	8522	A	14-8
8453	B	14-10	8488	A	2-6	8523	C	15-2
8454	B	14-10	8489	C	2-6	8524	B	15-2
8455	B	15-1	8490	B	2-7	8525	B	15-2
8456	C	14-10	8491	A	2-7	8526	A	15-3
8457	B	14-11	8492	C	14-2	8527	C	14-8
8458	A	14-11	8493	C	14-2	8528	C	15-3
8459	C	14-11	8494	C	14-3	8529	B	15-3
8460	A	14-2	8495	C	14-3	8530	C	15-3
8461	A	14-2	8496	B	14-3	8531	C	15-4
8462	C	14-11	8497	C	14-3	8532	B	15-4
8463	A	14-11	8498	B	14-3	8533	A	14-8
8464	C	15-1	8499	A	14-3	8534	C	14-8
8465	A	2-3	8500	C	14-4	8535	C	14-8
8466	A	2-1	8501	C	14-4	8536	A	15-4
8467	C	2-4	8502	C	14-4	8537	B	15-4